Solute Transport in Plants

TERTIARY LEVEL BIOLOGY

A series covering selected areas of biology at advanced undergraduate level. While designed specifically for course options at this level within universities and polytechnics, the series will be of great value to specialists and research workers in other fields who require knowledge of the essentials of a subject.

Recent titles in the series:

Social Behaviour in Mammals	Poole
Genetics of Microbes (2nd edn.)	Bainbridge
Seabird Ecology	Furness and Monaghan
The Biochemistry of Energy Utilization in Plants	Dennis
The Behavioural Ecology of Ants	Sudd and Franks
Anaerobic Bacteria	Holland, Knapp and Shoesmith
Biology of Fishes	Bone and Marshall
The Lichen-Forming Fungi	Hawksworth and Hill
An Introduction to Marine Science (2nd edn.)	Meadows and Campbell
Seed Dormancy and Germination	Bradbeer
Plant Growth Regulators	Roberts and Hooley
Plant Molecular Biology (2nd edn.)	Grierson and Covey
Polar Ecology	Stonehouse
The Estuarine Ecosystem (2nd edn.)	McLusky
Soil Biology	Wood
Photosynthesis	Gregory
The Cytoskeleton and Cell Motility	Preston, King and Hyams
Waterfowl Ecology	Owen and Black
Biology of Fresh Waters (2nd edn.)	Maitland
Tropical Rain Forest Ecology (2nd edn.)	Mabberley
Fish Ecology	Wootton

Tertiary Level Biology

Solute Transport in Plants

T.J. FLOWERS
Professor in Plant Physiology

and

A.R. YEO
Senior Research Fellow

School of Biological Sciences
at the
University of Sussex

BLACKIE ACADEMIC & PROFESSIONAL
An Imprint of Chapman & Hall
London · Glasgow · New York · Tokyo · Melbourne · Madras

Published by
Blackie Academic & Professional
An Imprint of Chapman & Hall
Wester Cleddens Road, Bishopbriggs, Glasgow G64 2NZ, UK

Chapman & Hall, 2–6 Boundary Row, London SE1 8HN, UK

Blackie Academic & Professional, Wester Cleddens Road, Bishopbriggs, Glasgow G64 2NZ, UK

Chapman & Hall, 29 West 35th Street, New York NY 10001–2291, USA

Chapman & Hall Japan, Thomson Publishing Japan, Hirakawacho Nemoto Building, 6F, 1-7-11 Hirakawa-cho, Chiyoda-ku, Tokyo 102, Japan

DA Book (Aust.) Pty Ltd., 648 Whitehorse Road, Mitcham 3132, Victoria, Australia

Chapman & Hall India, R. Seshadri, 32 Second Main Road, CIT East, Madras 600 035, India

---

First edition 1992

Typeset by Thomson Press (India) Limited, New Delhi

Printed in Great Britain by St Edmundsbury Press, Bury St Edmunds, Suffolk

ISBN 0 216 93221 1 (hbk)
0 216 93220 3 (pbk)
0 412 03271 6 (hbk) (USA)
0 412 03281 3 (pbk) (USA)

A catalogue record for this book is available from the British Library

Library of Congress Cataloguing-in-Publication data available

Preface

The study of solute transport in plants dates back to the beginnings of experimental plant physiology, but has its origins in the much earlier interests of humankind in agriculture. Given this lineage, it is not surprising that there have been many books on the transport of solutes in plants; texts on the closely related subject of mineral nutrition also commonly address the topic of ion transport. Why another book? Well, physiologists continue to make new discoveries. Particularly pertinent is the characterisation of enzymes that are able to transport protons across membranes during the hydrolysis of energy-rich bonds. These enzymes, which include the H^+-ATPases, are now known to be crucial for solute transport in plants and we have given them due emphasis. From an academic point of view, the transport systems in plants are now appreciated as worthy of study in their own right—not just as an extension of those systems already much more widely investigated in animals. From a wider perspective, understanding solute transport in plants is fundamental to understanding plants and the extent to which they can be manipulated for agricultural purposes.

As physiologists interested in the mechanisms of transport, we first set out in this book to examine the solutes in plants and where are they located. Our next consideration was to provide the tools by which solute movement can be understood: a vital part of this was to describe membranes and those enzymes catalysing transport. Our final aim was to deal with transport over increasing distance; from subcellular organelles through intercellular to long-distance transport. We have tried to summarise the literature to make it palatable for an undergraduate in Britain. By including references to books and major reviews we hope that we have also provided a starting point for those who want to take their studies further.

We would like to acknowledge the help of those who allowed us to use their data for figures and tables: Roger Discombe for his assistance with certain of the figures: Nasser Hajibagheri, John Hall, Roger Pearce, Rachel Leech and Julian Thorpe who provided us with micrographs. Our thanks go to Tony Moore and Mike Hasegawa who commented on particular chapters and especially to Sue Hitchings for reading and commenting on the whole text. Finally we thank Sam and Maggie who had to endure the whole process and another summer without a holiday.

TJF
ARY

Contents

Abbreviations and symbols

ABA	abscisic acid
ADP	adenosine diphosphate
ao	alternative oxidase
ATP	adenosine triphosphate
ATPase	adenosine triphosphatase
CAM	Crassulacean acid metabolism
C_a	carbon dioxide concentration in the air
C_i	intercellular carbon dioxide concentration
C_4	4-carbon compounds (hence C_4 plants)
DEAE	diethylaminoethyl
DCCD	dicyclohexylcarbodiimide
E	electrical potential
e_a	partial vapour pressure of water in the atmosphere
e_i	partial pressure of water vapour in intercellular spaces
ECS	extracellular space
F-type	a type of ATPase (F_0F_1-type)
FRI	fluorescence ratio imaging
FV	fast-vacuolar
G	gravitational potential
GTP	guanosine triphosphate
h	height in a gravitational field
I	current
J	flux
l	length
LAMMA	laser microprobe mass analyser
NADH	nicotinamide adenine dinucleotide
NADPH	nicotinamide adenine dinucleotide phosphate
NAD(P)H	NADH and NADPH
n_i	amount (moles) of substances (*i*)
n_j	amount (moles) of substances (*j*)
NMR	nuclear magnetic resonance
o	outside
P	pressure
P-type	a class of ATPase
PCMBS	p-chloromercuribenzene sulphonic acid
PEPC	phosphoenolpyruvate carboxylase
P_i	inorganic phosphate
PIXE	particle-induced X-ray emission
pmf	proton motive force
PPase	pyrophosphatase

PP_i	pyrophosphate
r	radius
Rubisco	ribulose bisphosphate carboxylase
SDS	sodium dodecylsulphate
T	temperature
TCA	tricarboxylic acid cycle
UQ	ubiquinone
V	voltage
V	vacuolar
V-type	a class of ATPase
VDACs	voltage-dependent anion-selective channels
x	xylem
η	viscosity
μ_j	chemical potential
π	osmotic potential (the negative of the osmotic pressure)
ψ_w	water potential

CHAPTER ONE
PLANTS AND THEIR SOLUTES

It is impossible to list all the solutes in plants: there are too many plants and too many solutes. Indeed it may even be difficult to know whether a particular substance is in solution in the plant or is complexed in some way that has reduced its solubility. The substance may, for example, be bound to a charged surface or have crystallised, or it may be part of a metalloprotein. It is easy enough, however, to extract from a plant a solution that can be analysed to determine the major solutes. By putting tissue under pressure, perhaps after freezing and thawing to disrupt the cellular structure, an extract—the plant sap—can be obtained. A vast number of such extracts have been made and their osmotic pressures (see section 2.8.2 for a definition of osmotic pressure) determined since de Vries began this type of work in the 1880s, although rather fewer have been analysed for their component solutes (see Stocking, 1956). Sugars, salts and organic acids account for the majority of the osmotic pressure of the sap (Figure 1.1A). Glucose, fructose and sucrose are the dominant sugars and potassium, calcium and magnesium the major cations: nitrate is generally the most important inorganic anion. The proportion of these solutes varies, however, with the particular part of the plant being analysed, its age and the habitat in which the plant is growing (see Kinzel, 1982). In fruits and in the swollen roots of beet, for example, sugars constitute the major osmotic solutes. There are other special cases where the sap is dominated by inorganic ions, particularly where plants are growing on saline soils (Table 1.1; Figure 1.1B). In yet other cases, organic acids such as malic acid may be present in high concentrations, particularly at the end of the night for those species using Crassulacean acid metabolism (see section 3.6).

1.1 Solutes in cells

The solutes in plants are invariably retained within cells. These cells may be undifferentiated, as in the meristematic tissues, or differentiated for

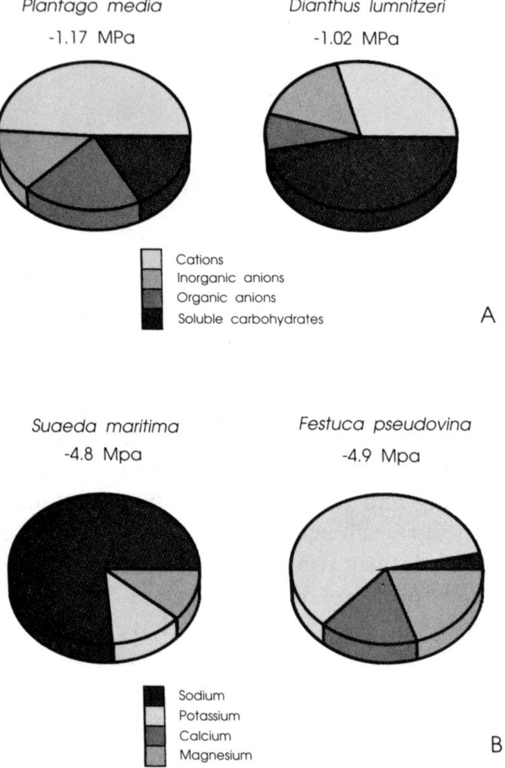

Figure 1.1 The proportion of the osmotic pressure of the cell sap contributed by various classes of solutes. (a) The contribution of different classes of solute in the cell sap of two dicoyledonous calcicoles (Redrawn from Kinzel, 1982). (b) Relative proportions of different cations in two contrasting halophytes. The total cation concentrations were 593 mol m^{-3} in *Suaeda* and 401 mol m^{-3} in *Festuca* (data of Albert and Kinzel (1973)). The total solute potential of the cell sap is shown underneath the species name.

processes such as storage or transport. Meristems are composed of small, approximately isodiametric, cells with a volume of between about 0.5 and 8 fm^3 (calculated as sphere or cube with radius or side between 10 and 20 μm; an fm^3 (femto m^3) or 10^{-15} m^3 is the same as a pl (pico litre) or 10^{-12} litre). Meristematic cells are densely cytoplasmic and contain few vacuoles (Figure 1.2a): the major component of their volume is the nucleus.

As development proceeds, cell volumes increase and reach values between 50 and 5000 fm^3 (Zimmerman and Steudle, 1980). The major

Table 1.1 Composition of the cell sap from the halophyte *Suaeda maritima* growing in a coastal salt marsh, calculated as a proportion of the osmotic potential (π) of the sap. (Data of Baumeister and Ernst (1978), taken from Kinzel (1982))

Component	Proportion of π (%)
Chloride	35.4
Sulphate	3.0
Phosphate	2.4
Organic acid	9.0
Sugar	1.9
Sodium	42.7
Potassium	3.1
Magnesium	2.5

Table 1.2 Volume fractions (as %) of mesophyll cells from the leaves of the halophyte *Suaeda maritima* grown in culture solution in the presence of 340 mol m^{-3} NaCl. Data of Hajibagheri (1984)

Cellular compartment	Relative volume (%)
Cell wall	4.3
Chloroplasts	12.7
Mitochondria	0.6
Cytoplasm	9.8
Vacuole	72.5

increase is in the volume of the vacuole(s). In mature cells, the vacuole is generally by far the largest compartment (Figure 1.2b; Table 1.2), although plastids may be an important feature of the volume of some cells.

The concentration of solutes in plant cells is greater than in the medium in which the roots are growing in all but exceptional cases, and usually by a considerable margin. There are two reasons for this. Firstly, plant organs that are not woody rely on hydrostatic pressure against a retaining cell wall to provide their support, shape and form. A high concentration of osmotically active solutes is needed to provide the driving force for the uptake of water to generate this hydrostatic pressure. Secondly, many

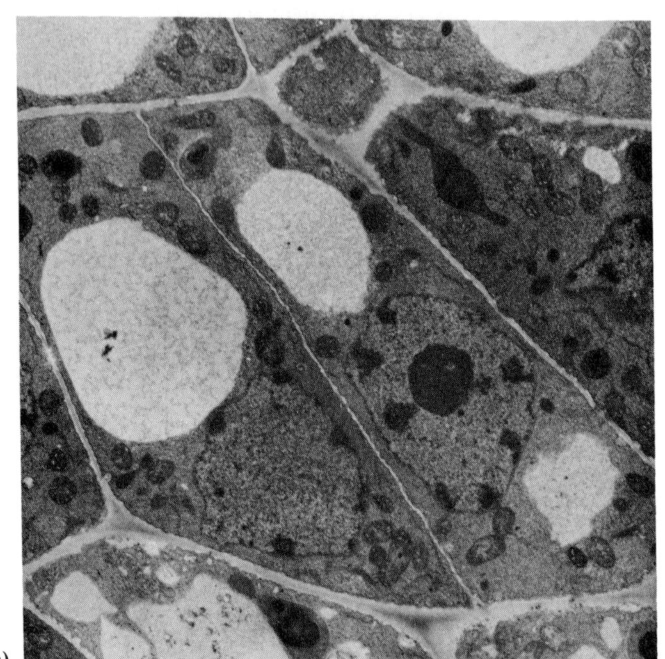
(a)

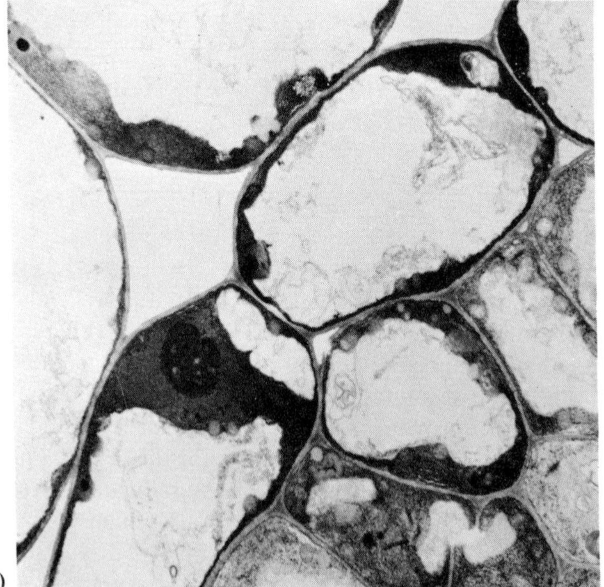
(b)

proteins require high (about 100 mol m^{-3}) concentrations of ions for their stability and activation. The latter function is inherently ion-specific and may be particularly so (for potassium) whilst the osmotic role can often be fulfilled by a variety of solutes (Flowers and Lächli, 1983).

1.2 Localisation of solutes within cells

Plant cells consist of many compartments: vacuole, cytoplasm, nucleus, mitochondria, plastids and other organelles within the cytoplasm. Solutes which are important within just one organelle, and so vital for the tissue as a whole, may have little impact upon the solute content of the cell. Consequently, the analysis of whole tissue provides only limited information and there is an ever-present need to know concentrations, and changes in concentration, within the smaller compartments. This analysis of cells and subcellular compartments, be it qualitative or quantitative, is problematical because of the small volumes involved. While there are a number of methods, varying greatly in resolution and in technical complexity, that can be used to try to analyse the solute concentration within subcellular compartments, none is of universal applicability; all have advantages and disadvantages. These methods are summarised below.

1.2.1 *Direct sampling*

Sap can be pressed from the plant and is primarily vacuolar in origin, because the vacuole is the major cellular compartment in mature cells. Normally, however, there is too much mixing of the cellular contents when sap is extracted for confidence to be placed in measurements made on such crude preparations. In some cases it is sufficient to analyse very small amounts of tissue, which are rather uniform in their degree of vacuolation, in order to deduce differences between vacuolar and cytoplasmic solute concentrations. The determinations rely on the very different degree of

---

Figure 1.2 Electron micrographs of meristematic and mature cells from the root of *Suaeda maritima*: (a) shows meristematic cells, the dominant features of each being their nucleus and the vacuoles (pale areas). The vacuoles occupy less than half of the volume of the cell. By way of contrast, the vacuole dominates the mature cells (b), the cytoplasm generally being parietal. The tissue was fixed in glutaraldehyde and post-stained with lead. (a) × 5400; (b) × 3000. The electron micrographs were kindly provided by Dr M.A. Hajibagheri of the Imperial Cancer Research Fund.

vacuolation, which must be measured or assumed, of meristematic and differentiated cells (see Table 1.3).

An alternative approach to the determination of compartmental solute concentrations is direct sampling and analysis. However, even though the vacuole is normally the largest compartment within a plant cell, only with certain algae that have giant cells (with a volume of about 10 μl; see Figure 2.12), is it *readily* possible to remove a sample of the vacuolar sap and analyse its contents. Attempts have been made to isolate vacuoles from normal plant cells and analyse them (Leigh, 1983 and Table 1.3), but there is often a considerable problem in obtaining a sufficient quantity of material for analysis and it is difficult to assess losses of solutes occurring during the isolation procedure. Recently, microtechniques have been developed that allow sampling, using a microcapillary, of the contents of individual vacuoles of normal plant cells. Great care has to be taken with the measurement of volume, since it is so small. Subsequently the osmotic

Table 1.3 Some recent examples of the methods that can be used to localise solutes within cells and tissues.

Method	Reference
Analysis of protoplasts and/or vacuoles	Leigh *et al.* (1981)
	Pahlich *et al.* (1983)
Analysis of small tissue samples	Jeschke and Stelter (1976)
	Hajibagheri *et al.* (1985)
Inference from biochemistry	Flowers *et al.* (1976)
Flux analysis	Hajibagheri *et al.* (1988)
Micro-autoradiography	Lüttge and Weigl (1965)
NMR	Ratcliffe (1965)
	Bental *et al.* (1988)
Microsampling (nl) and X-ray microanalysis	Malone *et al.* (1989)
Specific ion electrodes:	
K	Bowling (1987)
H	Reid and Smith (1988)
Ca	Felle (1989)
'Staining' e.g. with sodium cobaltinitrite	Harvey *et al.* (1979)
	Mentré and Escaig (1988)
Fluorescence ratio imaging	Tsien and Poenie (1986)
	Clarkson *et al.* (1988)
X-ray microanalysis:	
frozen hydrated (microprobe)	Huang and van Steveninck (1989)
freeze-substituted	Clipson *et al.* (1990)
Proton beam (micro-PIXE)	Makjanic *et al.* (1988)
Ion beam	Schaumann *et al.* (1988)
LAMMA	Heinrich (1989)

pressure of the sap can be determined from the freezing-point-depression caused by the dissolved solutes. This is measured in a tiny droplet of solution using a device called a *nanolitre osmometer*. Some of the individual elemental components can be determined by X-ray micro-analysis (see section 1.2.3 and Table 1.3 for references to recent papers using these methods).

Another technique that involves analysis on the microscale is the insertion into the cell of electrodes that are responsive to specific ions, for example potassium, hydrogen (protons), or calcium. The calcium and hydrogen electrodes can, in principle, be used to determine both ions in the cytoplasm and vacuoles, and the potassium electrode has been used for vacuoles and cell walls (Table 1.3). The advantages of using these electrodes are speed and applicability to intact tissue. The problems are knowing where the electrode tip is within the cell and limitations of sensitivity and specificity, though these are areas where there are continuing developments in technology.

1.2.2 *Indirect methods*

There are a variety of more or less indirect methods by which compartmental solute concentrations can be deduced. Some rely on the use of radionuclides, others on reactions with chemicals that are added to the tissue and yet others on physical or chemical properties of cellular solutes.

One means of deducing compartmental ion concentrations is to use so-called *flux*, or *compartmental, analysis*. Tissue is loaded with a radioactive element and its loss from the tissue followed with time. A three-compartments-in-series model of plant tissue (extracellular space consisting of cell walls and intercellular spaces, cytoplasm and vacuole) is simple to analyse mathematically and draws support from the countless cases where this has appeared to be an adequate description of the observed data (section 1.4.1). From such a model of the efflux, concentrations in the cytoplasm and cell walls and, under some circumstances, the concentration in the vacuole can be calculated.

If the tissue can indeed be loaded with a radioactive element, then this radionuclide can be located within the tissue using autoradiography and within cells using micro-autoradiography (in which the plant tissue is placed upon X-ray-sensitive film). The resolving power of the latter depends on the radionuclide; the more energetic the emission, the further the particles will travel before they interact with the photographic emulsion, and the more 'blurred' will be the image.

The concentration of some elements can also be deduced from their physical and chemical properties. Nuclear magnetic resonance (NMR) can be used in a few cases since the environment of an element shifts its NMR signal. For example, a phosphorus atom in the cytoplasm where the pH is close to neutrality will generate a different spectrum from a phosphorus atom sited in an acid vacuole, at a pH of perhaps 5.5 (see Ratcliffe, 1987). Circumstantial evidence of compartmental ion concentrations can also be obtained from the requirements of enzymes assayed *in vitro*. Many cytoplasmic enzymes show optimal activity at a hydrogen ion concentration of about 0.1 $mmol\,m^{-3}$ (pH 7), while protein synthesis, for example, exhibits optimal activity *in vitro* in 80–120 $mol\,m^{-3}$ potassium.

An alternative to the methods outlined so far is to use what might broadly be classed as stains, i.e. chemicals which when added to a tissue give an indication of the presence of a specific compound, and even its concentration, in conjunction with a light or electron microscope. An example is the use of sodium cobaltinitrite to stain for potassium. This was used many years ago with guard cells and has subsequently been used in a range of other cells (see Table 1.3). Recently, there has been considerable development of related techniques which can now be used on living cells to give quantitative estimates of certain elements. For example, the fluorescence excitation spectrum of the compound fura-2 changes depending on the concentration of calcium ions with which it is in contact (Figure 2.11). Therefore, if the fluorescence can be measured, the concentration of calcium can be determined from the ratio of the emission spectrum at two excitation wavelengths. The technique, known as fluorescence ratio imaging (FRI) (Tsien and Poenie, 1986), can be extremely sophisticated in providing tissue maps of the distribution of elements for which there are fluorescent probes (see Bush and Jones, 1990). The technique is limited, however, by the resolution of the light microscope to a few micrometres, and by the lack of specific probes for the majority of elements. It has been most successful in the measurements of calcium and hydrogen ions. Visualisation is also limited to single cells or to the surface layers of a tissue and requires that the dye remains in the compartment of interest (cf. Bush and Jones, 1990). If ions are to be located within the cytoplasm, with a width which may be only 500 nm, then techniques with greater resolving power are necessary.

1.2.3 *X-ray microanalysis*

The most common method widely available for increasing the resolution of microscopy is the use of electrons in place of visible light. Because the wavelength of electrons is much shorter than that of light, the resolving power of an electron beam is much greater than that of a beam of light. As far as elemental analysis is concerned, the electron microscope is particularly valuable as it is easily (if not cheaply) adapted to become an analytical instrument, at least for elements with an atomic number of 11 (sodium) and above (Hall and Gupta, 1983). When elements are bombarded with high energy electrons, they emit X-rays that can be analysed in terms of their energy or wavelength. Both these properties are specific to given elements and the response can be quantified, generating a powerful analytical tool. The energy of the X-ray is broadly proportional to atomic number for most elements of biological interest (it is not generally possible to detect hydrogen, carbon, oxygen or nitrogen by this means).

Aside from the inability to determine the lighter elements, the chief problems for the application of X-ray microanalysis to biological material lie in the preparation of the tissue prior to analysis and the inability to distinguish soluble from insoluble forms of an element whose concentration is being determined. Electron microscopes require relatively high vacuum as electrons have very poor powers of penetration through gases such as air or water vapour. It is possible to use hydrated tissue only if it is frozen to minimise the vapour pressure of water in the microscope and the microscope must be used in the 'scanning' mode. In this mode, electrons emitted from the surface of the specimen are used to generate the image because those bombarding the tissue cannot pass through the ice and generate an image, as occurs when the microscope is used in the 'transmission' mode. Used in the scanning mode for analysis of frozen specimens, the electron microscope has been dubbed the 'X-ray microprobe' (see Duncan, 1990 and Table 1.3). Under most circumstances, the resolution is still below that required to localise elements within the cytoplasm. Visualisation of surface detail has often been poor and it is difficult to produce a fracture-surface which is 'flat' enough not to cause artefacts in the analysis. For resolution within the cytoplasm to be reliably achieved it is necessary to use thin sections. However, the preparation of thin sections requires dehydration and embedding with a plastic resin of some form or other. Unless specific precautions are taken, this results in the loss of ions from the tissues. Consequently, a specific preparative method has to be adopted.

One such method is known as freeze substitution (see Flowers and Lächli, 1983). In this procedure, the tissue is first frozen as rapidly as is practicable by immersion in molten nitrogen. The tissue is then dehydrated while frozen (at below $-30°C$) by exchanging the water molecules for those of an organic solvent that is miscible with the resin (a solvent such as acetone, diethyl ether or ethanol). The solvent must not, of course, dissolve any ion of interest. The solvent and tissue can then be warmed and infiltrated with resin which is polymerised in the normal manner. The final difficulty is that the sections cannot be cut onto water as is routine when sectioning for electron microscopy (at least not with the resins in general use, because they are quite permeable to water) but have to be cut with a dry (glass) knife. Subsequently, however, they can be used for analysis with a resolution of a few nanometres, sufficient to resolve subcellular compartments within the cytoplasm. Because conventional electron-dense stains (reliant upon the heavy metals osmium, uranium and lead) are applied in aqueous solutions, sections for X-ray microanalysis must remain unstained and consequently the image contrast is poor (compare Figures 1.2 and 1.3). The analysis determines the amount of an element within the volume of the specimen that is analysed: it does not determine whether that element was in solution in the tissue or present in some insoluble form. Whether or not the element was in solution must be argued from the nature of the compartment, the concentration of the element or the presence or absence of other elements likely to form insoluble complexes.

There are alternatives to the use of electrons and X-rays for the analysis of biological material. Methods based on ion and proton beams (micro-PIXE) and on lasers (LAMMA; a laser ionises material located under an optical microscope and the vaporised material is subjected to mass spectrometry) have been developed, but are not in widespread use as the equipment is highly specialised and not widely available.

1.3 Solute concentrations and ion activities

Although it is frequently possible to localise a particular element, it is much more difficult to deduce its concentration on the basis of the water content of the tissue, let alone its chemical activity. Chemical activity is the *effective* concentration of a solute in solution; effective since interactions between solute molecules, for example, can alter their behaviour in solution. It is the chemical activity that governs the reactivity and movement of any particular ion. Estimation of activity is not so much a problem for solutes in vacuoles (because the vacuolar sap is generally considered to be a simple solution, though it would be a mistake to consider it an *ideal* solution because the concentrations are too high), but deducing concentrations and activities in cytoplasm and cell walls is a real difficulty. The problems arise from uncertainty of the water content of these compartments and the degree

Table 1.4 Estimates of ion concentrations in subcellular compartments of plant cells made from data in the literature. Plants are assumed to be growing under normal nutrition and not suffering from adverse environmental conditions. Note the variation in the units of concentration

Compartment	Ion	Concentration
Cytoplasm	H	$\approx 32\,\mu mol\,m^{-3}$
	Ca	$30–400\,\mu mol\,m^{-3}$
	K	$100–200\,mol\,m^{-3}$
	P	$5–10\,mol\,m^{-3}$
Chloroplast	Cl	$100\,mol\,m^{-3}$
Vacuole	H	$\approx 3\,mmol\,m^{-3}$
	K	$20–100\,mol\,m^{-3}$
	Ca	$>1\,mol\,m^{-3}$
	P	$5–15\,mol\,m^{-3}$
Cell wall	K	$300–500\,mol\,m^{-3}$

to which ions are influenced by other charges. The consensus seems to be that 70 to 90% of the water in the cytoplasm behaves in a similar way to pure water, while the remainder is influenced significantly by proteins (see Wyn Jones and Pollard, 1983). Close to charged surfaces water molecules become 'organised', comparable to the crystal lattices of ice and, like ice, organised water does not have the remarkable solvent properties associated with liquid water. The result is that true ion concentrations are higher than might be calculated from estimates of total volume or water content. Some estimates of ion concentrations in various subcellular compartments are given in Table 1.4. Activities generally remain unknown, except where they have been measured using microelectrodes. Since activity coefficients (the amount by which the effective concentration differs from the measured concentration; the activity coefficient of $200\,mol\,m^{-3}$ KCl is 0.718) of biologically important ions are far from unity at physiological concentrations, differences between activities and concentrations may be substantial.

1.3.1 *Localisation of organic solutes*

The discussion so far has centred on the localisation of inorganic, rather than organic solutes. While the localisation of these inorganic solutes is not easy, the visualisation of organic solutes presents even greater problems. Elemental analysis conveys little information when all the compounds of interest consist mostly of H, C, N and O. Furthermore, the great variety of possible compounds makes the analysis of specific molecules difficult. Specific compounds must be made to take part in a unique reaction in which they combine in some way with an identifiable molecule or stain.

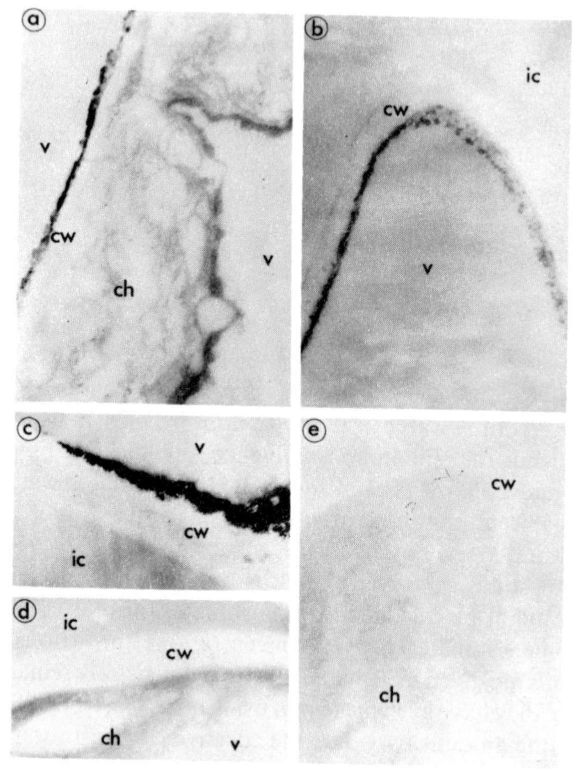

Figure 1.3 The localisation of an organic solute, glycinebetaine, in cells from the leaf of *Suaeda maritima* and its absence from *Oryza sativa*. The tissues were freeze substituted in the presence of iodoplatinic acid which complexes with glycinebetaine to produce an electron opaque deposit. The amount of glycinebetaine present in the tissue and hence the amount of the deposit, increases as the amount of sodium chloride present in the medium in which *Suaeda maritima* is growing is increased. (a) and (b), *S. maritima* grown in the presence of 170 mol m^{-3} sodium chloride showing deposits of platinum, marking the presence of its complex with glycinebetaine in the cytoplasm and no staining in the vacuole. (c) *S. maritima* grown in the presence of 510 mol m^{-3} sodium chloride showing a particularly dense deposit. There is little evidence of the presence of glycinebetaine in plants grown in tap water alone (d) or in rice (e). Magnification: (a) ×15 200; (b) ×22 700; (c) ×18 200; (d) ×12 500. The micrograph was kindly provided by Professor J.L. Hall, University of Southampton. Reproduced from Hall *et al.* (1978) with permission.

Glycinebetaine has, for example, been localised primarily in the cytoplasm of the cells of the halophyte *Suaeda maritima* by reacting it with iodoplatinic acid under conditions in which the solute was retained in the tissue by preparation using freeze substitution (Figure 1.3). However, such procedures are rarely possible. A very specific (bio)chemical reaction with an electron-dense compound (in this example the high atomic number element Pt) is required.

The localisation and quantity of organic solutes can, under some circumstances, be determined by NMR. However, this is an average over numerous cells because a fairly large sample is needed to generate the NMR spectrum.

1.4 Conclusions

For the analysis of solutes in plant material there can be no generalisation about the technique of choice. Each technique has advantages, disadvantages, benefits and limitations according to the element of interest, its concentration, the tissue and the type of experiment. As an example, X-ray microanalysis has an unsurpassed spatial resolution, but it cannot measure intracellular pH (H cannot be detected) or cytoplasmic calcium (the concentration is too low) and it cannot be used on 'living' cells (because of the requirements of the electron microscope).

Whatever the solute, however, it is contained within a compartment or compartments in the plant. These compartments are normally cells or parts of cells and are mostly delimited by membranes. The properties of the membranes are central to the understanding of solute transport in plants and we shall examine their properties and how they influence the movement of solutes in the next chapter.

1.5 Addendum

1.5.1 *Flux analysis*

A plant cell can be described by a compartmental model with two major resistances to the movement of ions, the plasmalemma and the tonoplast. Ions moving between the outside and the vacuole must cross both barriers, the characteristics of which are important in cellular function. Flux analysis allows the determination of the characteristics of these membranes and the compartments they define. It is based on tracing movement of ions, with radionuclides, in a system which is otherwise at flux equilibrium. Tissue is loaded with a radioactive tracer of a given external concentration and the efflux of the tracer is followed after it has been removed from the external solution. Only the tracer is removed: the chemical concentration of the element in the external solution remains essentially unchanged.

In practical terms, tissue is excised and placed in radioisotope at a known specific activity for a period of hours. It is then removed from the radiolabel and transferred to unlabelled solution. This is changed (by transferring the tissue) after (say) 1, 2, 4, 8, 15, 30, 45 and 60 min, and 1.5, 2, 2.5 and 3 h.

If the data fit the three-compartment model (extracellular space, ECS, cytoplasm, C, and vacuole, V) then the efflux of radioisotope should be composed of three first-order rates of loss of activity (from ECS, C and V respectively), but superimposed on one another. First order means that the rate of efflux is proportional to the concentration, i.e.

$$\frac{\partial c}{\partial t} = -kc$$

where t is time, c is the concentration in the tissue and k is the proportionality constant. On rearranging,

$$\frac{\partial c}{c} = -k\partial t$$

which may be integrated

$$\int \frac{\partial c}{c} = -\int k\partial t$$

as

$$\int \frac{\partial c}{c} = \ln c = -\int k\partial t = -kt + z$$

Where z is a constant from the integration. Thus a plot of the logarithm of the concentration of ions in the tissue against t should be a straight line.

These calculations are used to determine the efflux with time, the amount of isotope remaining in the tissue after efflux and to plot the logarithm of the activity remaining in the tissue against t. A straight line is fitted to the linear portion of the curve. This represents the exponential rate constant for efflux from the vacuole (Figure 1.4A). Before the days of computers and calculators it was easier to work with logarithms to the base 10 rather than base e, a convention kept in the example ($\ln n = 2.303 \log n$).

In order to resolve the other two phases (ECS and C), the vacuolar efflux must be subtracted (it is assumed that efflux from all three compartments takes place simultaneously from $t = 0$). This can be done since we now know the relationship for the efflux from the vacuole with time. The straight line representing this relationship is extrapolated to t_0, the logarithm of the counts from the vacuolar phase determined for each time, antilogged and subtracted from the counts, and the data replotted (Figure 1.4B). The procedure is repeated for the final phase of efflux, i.e. from the extracellular spaces which include the cell wall (Figure 1.4C). Note that the timescales

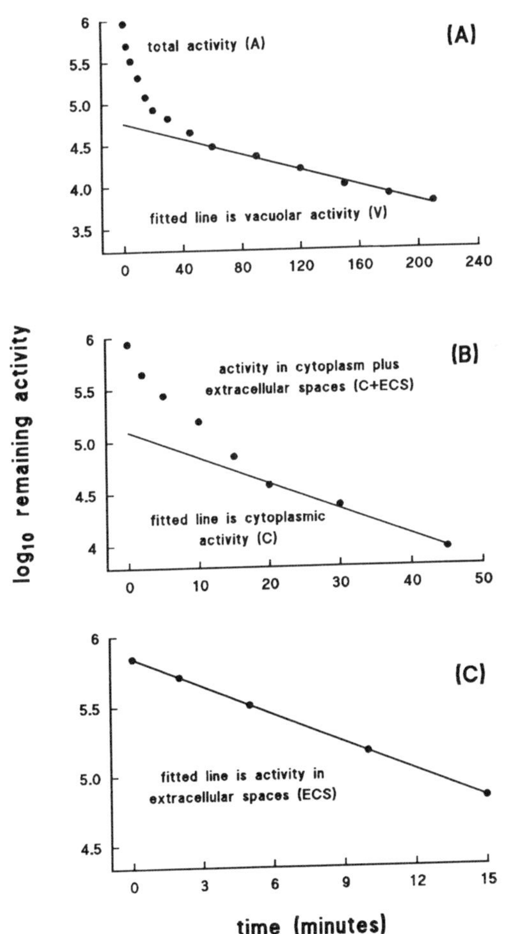

Figure 1.4 Flux analysis. The resolution of efflux from three compartments. See text for details.

(the x-axes) differ, and the rate constants for the three phases are very different. The rate constant for the cytoplasmic phase (exchange across the plasma membrane) is greater than for the vacuolar phase (exchange across the tonoplast). The rate constant for the extracellular spaces, which does not involve exchange across a membrane, is greater still. Rates of exchange are commonly described by half-times.

CHAPTER TWO

MEMBRANES AND THE STUDY OF TRANSPORT

2.1 Membrane structure

By the end of the last century, it was clear that there was a discrete membrane surrounding cells and that it had a selective permeability to solutes. By the end of the first quarter of this century, it had been established that the amount of lipid present in human erythrocytes was sufficient to cover twice the surface area of the cell, from which it was concluded that the cell membrane or *plasma membrane* must be composed of a lipid bilayer. Ten years on, it was found that the surface tension of fish and sea urchin eggs in water was very different from that of lipid droplets in water and the deduction made that the plasma membrane was not composed of lipid alone, but also contained protein. However, it was not until the advent of the electron microscope that membranes could be seen: they appeared electron dense in thin sections. In many instances two lines separated by a space were visible (cf. Figure 2.1a). As the technology associated with electron microscopy developed, it became possible not only to view surface features of frozen specimens, but also to fracture frozen material thus exposing the inner surfaces of cells. In micrographs (post-1957) of membranes that had been freeze-fractured, a granular matrix with embedded particles was generally visible (Figure 2.1b). These pictures were particularly useful as the membrane had not been treated chemically before being viewed with the electron microscope: the cell was simply frozen, fractured and some of the ice etched away. During fracturing the membrane split down its centre (the two halves of the bilayer separate) presenting a view of the inner faces of the membrane. The pictures suggest that some particles are not simply adhering to the membrane surface, but are an integral part of the membrane.

In working out the structure of membranes, images from the electron microscope have to be interpreted in the knowledge of the chemical components of the membrane. Chemical analysis has now shown membranes to be composed largely of lipids and proteins, with some

(a)

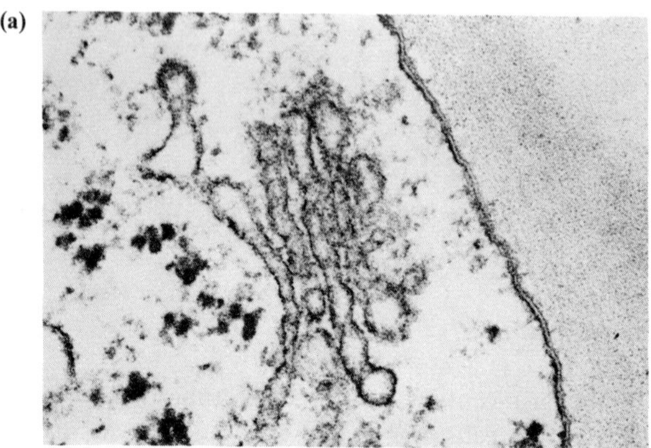

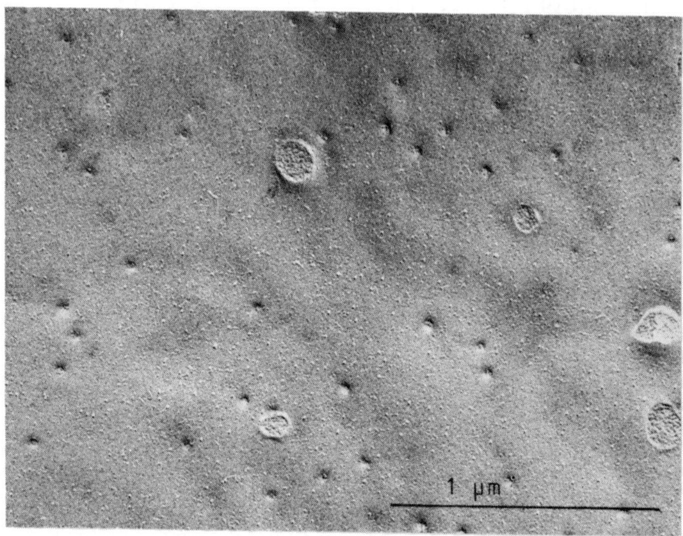

(b)

Figure 2.1 Electron micrographs of plant plasma membranes. (a) Part of a xylem parenchyma cell from the root of *Zea mays*, taken 25 cm from the root tip. The tissue was fixed in glutaraldehyde and stained in osmium with post-staining in zinc uranyl acetate. The micrograph shows the plasma membrane as a double line lying between the cell wall and the cytoplasm, which in this case is dominated by a Golgi body. (b) A freeze-fractured view of the exoplasmic face of the plasma membrane of a wheat leaf base meristematic cell. The membrane splits so that the micrograph is of an inner surface and the cell wall lies behind the membrane. There are particles embedded in the membrane, intermediate-sized indentations which are plasmodesmata and five larger features that are thought to be due to vesicle fusion. (a) × 18 000 The micrograph of the fractured membrane was kindly provided by Dr R. Pearce, University of Newcastle upon Tyne.

carbohydrate. The lipids are chiefly phospholipids, sphingolipids, glycolipids and sterols. Figure 2.2 shows the structure of phosphatidyl choline illustrating the hydrophilic head group which faces outwards into the aqueous phase, and the hydrophobic tail region which contributes to the core of the membrane. The proteins vary, depending upon the origin and function of the particular membrane under consideration. The carbohydrate is attached to lipid or protein, as glycolipid or glycoprotein.

There have been a number of landmarks in the deduction of membrane structure. A particularly famous model was due to Danielli and Davson who, in the mid-1930s, suggested that membranes consisted of a lipid bilayer covered by a layer of protein on either side. The two dark lines in electron micrographs were equated with the zone in which protein and lipid interacted. In the late 1950s, Robertson modified this model and, since all

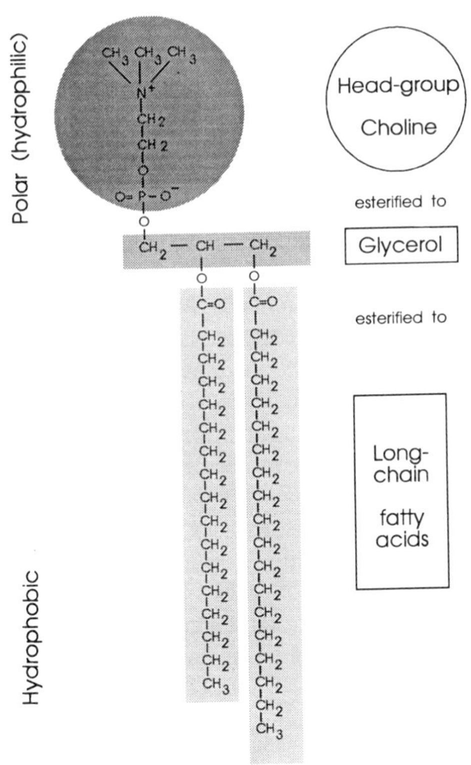

Figure 2.2 The structure of phosphatidyl choline, a typical bilayer-forming membrane lipid.

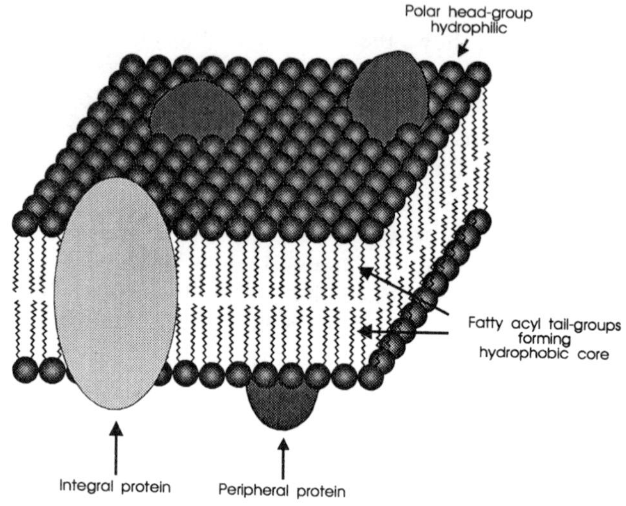

Figure 2.3 A biological membrane drawn according to the fluid mosaic model of Singer and Nicholson.

membranes appeared similar in electron micrographs, he developed the concept of the 'unit membrane'; again a lipid bilayer, but with protein in an extended, rather than a globular form and an allowance made for sidedness. Membranes may separate very different compartments and it is reasonable that they have an outside and an inside. By the early 1970s the evidence for and against various model structures was sufficient to allow Singer and Nicholson to draw together the information in the so-called 'fluid mosaic model' (Figure 2.3) which is still acceptable today. The components of the membrane are hypothesised to be arranged as a lipid bilayer in equilibrium with an array of globular protein molecules. The proteins may be peripheral (*extrinsic*) to the membrane and easily detached, or an integral part of the membrane structure (*intrinsic* proteins). The lipid molecules, under normal conditions, constitute a fluid (liquid crystal) phase in which the proteins are relatively free to migrate and, of crucial importance with respect to their transport properties, to change conformation. By analogy, the proteins may be considered to be dissolved in the liquid crystal.

2.2 Membranes of plant cells

Like the cells of other eukaryotes, plant cells are divided up by membranes. However, the outermost membrane is not the outer face that the plant presents to its environment. This is the cell wall (see also section 5.1.6). Behind this wall, which is a vitally important complex chiefly composed of carbohydrates, lies the plasma membrane or *plasmalemma* (we will use the term 'plasma membrane' throughout this text), the membrane forming the outer boundary of the *protoplast*. Within the plasma membrane lies the cytoplasm containing the various cellular organelles, each membrane-bound (the nucleus, the chloroplasts, the microbodies and mitochondria), and other membranes such as those of the endoplasmic reticulum and the Golgi apparatus. In mature cells, the dominant feature is the central vacuole (Figure 1.2B) which is the site of most of the solutes in these cells (Table 1.2). The vacuole is bounded by a membrane known as the *tonoplast*. Although many, or perhaps all, of the membranes play some role in solute transport within and between cells, it is the tonoplast that governs solute storage in the central vacuole, and the outer plasma membrane that first interacts with solutes in their passage into the cell. The plasma membrane is particularly interesting, since it does not simply surround an individual cell, but has connections through tiny, but specialised, pores to neighbouring cells (the *plasmodesmata*, see section 4.4.2). In this sense, the cytoplasm of plants is a living continuity, whilst the vacuoles are isolated from one another.

2.3 Obtaining membranes for transport studies

For most animal cells the plasma membrane is essentially the outer face of the cell, so it is relatively accessible for experiments. Not so for plants: the plant cell wall is a tough barrier modulating the access of external solutes to the plasma membrane. Because the cell wall has a large ion-exchange capacity and because it is a substantial unstirred layer (see Figure 2.7), it intervenes both chemically and physically between the outside solution and the surface of the plasma membrane. In plants the plasma membrane is usually adpressed tightly against the cell wall by positive hydrostatic pressure. Furthermore, mature tissues contain cells which have developed rigid and/or impregnable properties as part of their cell walls. Thus, although much physiological information has been obtained about ion transport in plants from studies with intact tissues or cells (see chapter 5), knowledge of the role of the plasma membrane itself is generally confounded by the effects of the cell wall. Other cellular membranes are similarly inaccessible to the experimenter as they are situated within the cell. Consequently, there have been many attempts to isolate the cell's membranes, so that they may be studied independently.

2.3.1 *Homogenisation and centrifugation*

One of the simplest approaches to the preparation of subcellular fractions has been to homogenise tissues and subject the homogenates to differential centrifugation. This allows organelles to be separated and is the basis of much modern biochemistry. A similar procedure may also be employed to prepare fractions that contain membranes, rather than intact organelles (Figure 2.4). Cellular homogenates are cleared of major organelles such as

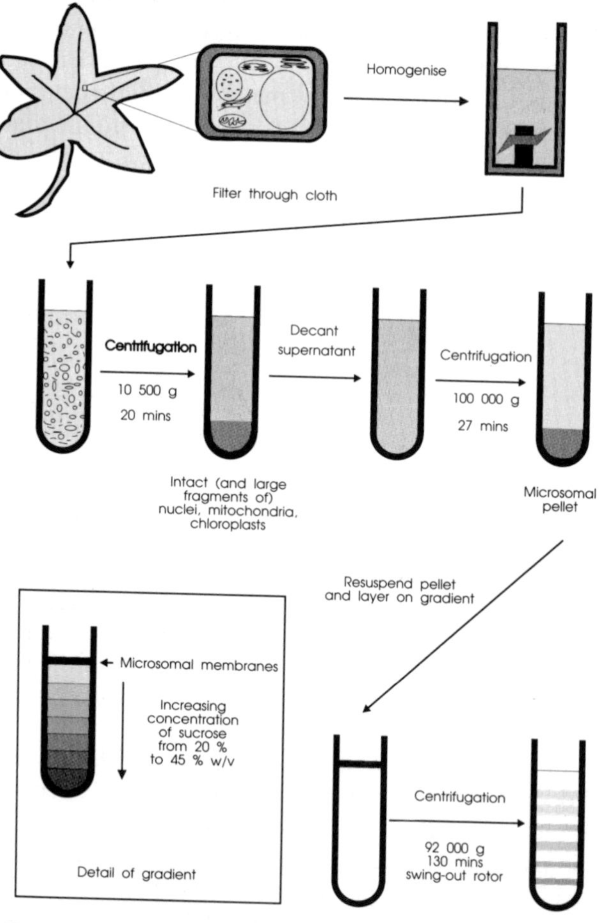

Figure 2.4 Differential and density gradient centrifugation. The figure shows the procedure, beginning with the homogenisation of a leaf, for the preparation of membrane fractions by differential and density gradient centrifugation.

plastids and mitochondria by low-speed centrifugation at about 10 000 × g for 15 minutes, which is sufficient to pellet these dense structures. The remaining membranes are then sedimented at higher speed (80 000 to 100 000 × g for about 60 minutes; but see Hall, 1983). The only remaining problem is the separation of the individual membranes.

In principle, the preparation of membrane fractions is a relatively simple matter. In practice, the major problem is that the shearing forces needed to disrupt the tough cell wall fragment not only the plasma membrane, tonoplast and internal membranes of the endoplasmic reticulum, but also the cytoplasmic organelles, including the mitochondria and plastids. The net result is that the homogenisation produces a mêlée of membrane fragments constituting the so-called 'microsomal' fraction, which then

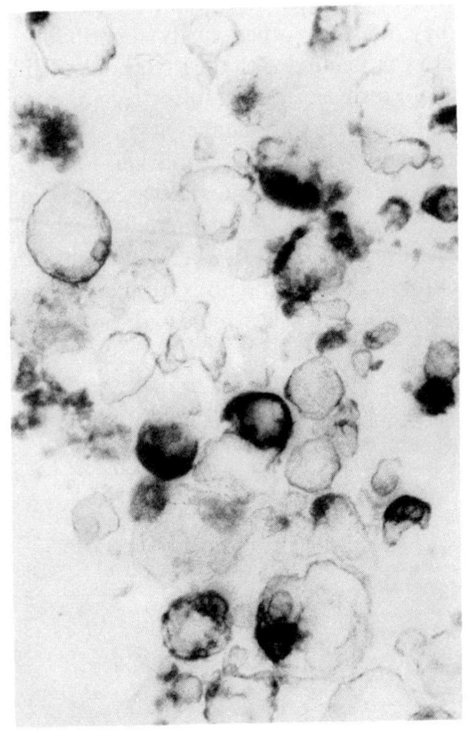

Figure 2.5 An electron micrograph of vesicles of a fraction enriched in tonoplast membranes prepared from the shoots of the halophyte *Suaeda maritima*. The micrograph was provided by Dr J. Thorpe, University of Sussex and data obtained using a similar preparation are shown in Figure 3.4. Magnification ×40 800.

requires density gradient centrifugation if individual membranes (e.g. plasma membrane and tonoplast) are to be separated. The microsomal pellet must be resuspended in a suitable buffer and placed over a density gradient, commonly a discontinuous gradient of sucrose (with steps ranging from 5% to 45% w/v; Figure 2.4). Membrane fractions, which are in the form of small vesicles (about 300 nm in diameter; Figure 2.5) separate at their equilibrium densities (Table 2.1) after centrifugation at about 80 000 × g for 150 minutes.

Once fractionated, an added difficulty is to identify from where each of the membranes originated: is it plasma membrane, endoplasmic reticulum or tonoplast, for example? As with all subcellular fractionations, the identification of particular components relies heavily on the use of 'markers'. These markers may be enzymes that are known to be associated only with a particular organelle (e.g. cytochrome oxidase and the mitochondrion) or they may be other easily identifiable compounds (e.g. chlorophyll, which is associated with the thylakoids; Table 2.1). It is only in the last ten years, however, that results from methods of membrane preparation other than density gradient centrifugation of cell homogenates (see below) have brought to light enzyme markers for the plasma membrane and tonoplast: two adenosine triphosphatases and a pyrophosphatase (Table 2.1; see also section 3.2). Unfortunately much of the earlier work, prior to the 1980s, was hampered by the lack of this information.

Analysis of membranes in the fractions from density gradients prepared from homogenates of plant tissues indicates that these fractions do not

Table 2.1 Some characteristics of membranes isolated from plant cells. The data are taken from Hall and Moore (1983), Hall *et al.* (1982) and Robards (1970)

Membrane	Equilibrium density (g ml^{-1})	Thickness in electron micrographs (nm)	Marker
Endoplasmic reticulum	1.08–1.13	7.5	Cytochrome c reductase
Golgi apparatus	1.135–1.15	6	Inosine diphosphatase
Plasma membrane	1.13–1.20	10	ATPase[a]
Tonoplast	1.06–1.12	7	Pyrophosphatase ATPase[b]

[a]vanadate sensitive (see section 3.2.2.1)
[b]vanadate insensitive but nitrate sensitive (see section 3.2.2.2)

contain just one type of membrane. The fractions may be enriched in a particular membrane, but they are far from pure. Individual membranes are mixed during homogenisation and centrifugation and this presents the opportunity for membranes from different parts of the cell to coalesce. Furthermore, the position of a vesicle in a density gradient may be modified by inclusions that come from different membranes and by material adsorbed onto the surface. The importance of this cross-contamination varies according to the use to which the membrane preparations are to be put. If the purpose of the fractionation is only to increase the specific activity of a component which independent evidence assigns to a particular membrane, then the procedure is perfectly serviceable. If the purpose is a fact-finding analysis of the composition of a particular membrane system then a density-gradient separation of a whole-tissue homogenate is not a promising starting material—the membrane fraction is simply not pure enough.

2.3.2 *Tonoplast*

The vacuoles are surrounded by a membrane with a low specific gravity and fractions enriched in tonoplast can be prepared by density gradient centrifugation where the membranes reach equilibrium at specific gravities of 1.06 to 1.12 $g\,ml^{-1}$ on sucrose gradients. However, the limitations described in the previous section have lead to the search for other methods by which tonoplast might be prepared. For the vacuolar membrane, mechanical slicing of tissue, which may or may not have been previously plasmolysed, has been very successful in specific cases (see Wagner, 1983). The benefit of plasmolysis is that it frees the protoplast from intimate contact with the cell wall and gives it a more uniform, spheroid shape. Slicing releases vacuoles that can then be collected by low speed centrifugation. Particularly successful has been the use of red beetroot as not only can large amounts of tissue be obtained easily (an important factor as the efficiency of the method is low) but the vacuoles are readily identified by the presence of the red pigment, betacyanin.

The release of vacuoles from cells was an important observation, but not all tissues can easily be obtained in bulk or are amenable to mechanical slicing. For these more difficult tissues vacuoles can be released from protoplasts which are prepared by digesting the cell walls enzymatically (Figure 2.6; Wagner, 1983): the vacuoles are released by osmotic lysis, which may be aided by the use of phosphate buffers, by some form of shear force or by treatment with the polybase, DEAE-dextran (Wagner, 1983). Vacuoles

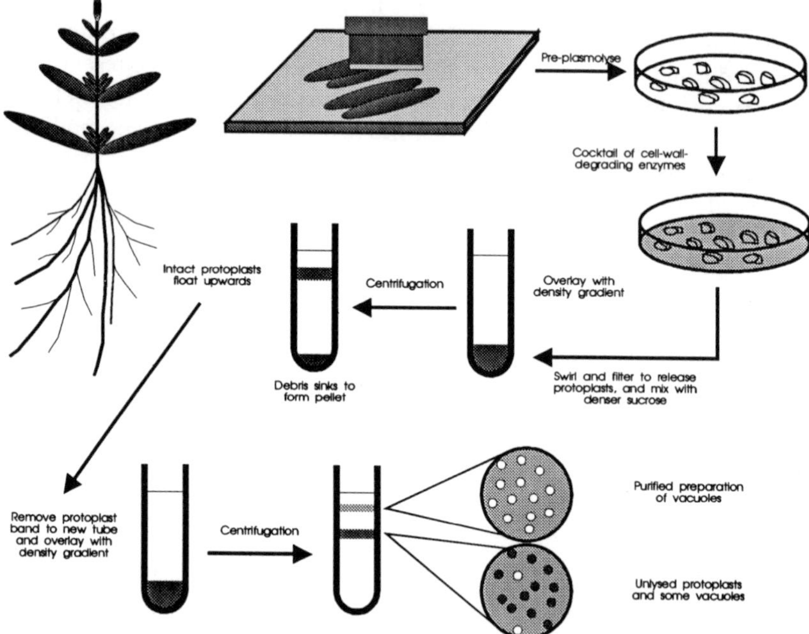

Figure 2.6 The preparation of vacuoles from protoplasts. Tissue is first chopped with a razor blade, then incubated in a cocktail of enzymes. Protoplasts and vacuoles are separated by centrifugation.

can be collected by sedimentation or floatation. Such procedures played a key role in the identification of two marker enzymes for the tonoplast; a vanadate-insensitive adenosine triphosphatase (ATPase) and a pyrophosphatase (see also section 3.2).

2.3.3 *Plasma membrane*

It is difficult to separate plasma membrane from the other parts of the cell, so its purification is not easy. Conventionally, the membrane, which is denser than tonoplast, has been obtained after density gradient centrifugation from a fraction with a specific gravity between 1.13 and 1.20 g ml^{-1}. It is characterised by the presence of a vanadate- sensitive ATPase (see section 3.2.2.1). Recently, purification procedures have utilised the fact that membranes can be partitioned between two phases composed of aqueous solutions of polymers. The method relies on the mixing of two water-soluble polymers in concentrations such that the two phases separate (Larsson, 1983). Suspended particles partition between the two phases according to surface charge and other less certain properties: plasma membrane partitions into the upper phase of a dextran/polyethylene glycol system. Such membrane preparations have been reported to be purer than those obtained by density gradient centrifugation alone (see Sandelius and Morré (1990) for details of methods available for the preparation of plasma membrane from plants), but even membranes obtained from phase partitioning remain 'enriched' rather than pure.

These recent developments have continued to rely on homogenisation of tissues to provide the initial material. Protoplasts (cf. Figure 2.6) would, however, appear to be underexploited as a starting material. The small shear forces needed to disrupt protoplasts means that differential centrifugation removes organelles with much greater efficiency and with much less contamination than when applied to tissue homogenates. Indeed, protoplasts are often the material of choice for obtaining intact chloroplasts for photosynthetic studies. One possible drawback to the use of protoplasts as a source of plasma membrane is that the outer surface of the membrane will have been exposed to the cocktail of cell-wall-degrading enzymes.

2.4 Driving forces for movement

Substances move for a variety of reasons: they may diffuse or work may be done to bring about movement. Here we will concentrate on diffusion, leaving movement requiring work for later sections of the book. Diffusion comes about through random thermal movements of substances and, at least for neutral molecules, results in the levelling out of concentration differences in adjacent regions, providing, of course, there are no impermeable barriers to the movement of the substances. The process of diffusion is particularly important in studying solute transport as it is one of the means by which solutes reach membranes and cross them.

The process of diffusion was studied in the middle of the last century and described quantitatively by Fick, who gave his name to laws describing the process (Fick's laws of diffusion). An important conclusion was that the *flux* of a substance j, i.e. the amount of j passing through a unit area in a unit time (also known as the *flux density*), is related to its concentration (c_j) gradient (the rate of change of c_j with distance, x, i.e. $\partial c_j/\partial x$) by the equation:

$$J_j = -D_j(\partial c/\partial x)$$

where the flux, J_j, is in mol m^{-2} s^{-1}, the concentration, c_j, is in mol m^{-3} and the distance, x, is in m. The negative sign indicates that the diffusion is away from the region of high concentration and towards a region of low concentration (the driving force is $-\partial c_j/\partial x$). The coefficient of proportionality, D_j, is known as the diffusion coefficient: it has dimensions m^2 s^{-1}. It is not a constant even in a simple solution as it varies with concentration and temperature. Typical values of diffusion coefficients are shown in Table 2.2.

Further analysis of diffusion is mathematically complex, but certain simplifications allow the calculation of the time taken for a proportion of the molecules to diffuse a certain distance from their origin. For example, the time for 50% of an original population of molecules ($t_{0.5}$) to diffuse a given distance ($x_{0.5}$) is given by the equation (Nobel, 1983):

$$x_{0.5}^2 = 2.8 D_j t_{0.5}$$

It is interesting to calculate the time taken for half a population of molecules to cross the distance equivalent to a cell (of 50 μm) or to travel the height of a small plant (500 mm). In the case of the cell, this time ($t_{0.5}$) equals 0.9 s, while in the case of the plant it is 2 years. On the time-scale on which living cells operate, diffusion is a perfectly adequate transport process for small distances but will not suffice over longer ones.

Apart from distance, membranes are also a limitation to diffusion. Diffusion coefficients of small solutes in membranes are often 10^6 times lower than in an aqueous solution. Again, as for diffusion through a solution, the driving force is the negative gradient to the concentration

Table 2.2 Diffusion coefficients for some common solutes at varying temperatures and concentrations. The data are taken from Robinson Stokes (1959) and Weast (1986)

Solute	Concentration	Diffusion coefficient (m^2 s^{-1})	Temperature (°C)
KCl	5 mol m^{-3}	1.71×10^{-9}	20
KCl	5 mol m^{-3}	1.93×10^{-9}	25
KCl	5 mol m^{-3}	2.16×10^{-9}	30
KCl	10 mol m^{-3}	1.92×10^{-9}	25
NaCl	5 mol m^{-3}	1.56×10^{-9}	25
$CaCl_2$	5 mol m^{-3}	1.21×10^{-9}	25
Glucose	22 mol m^{-3}	0.67×10^{-9}	25
Sucrose	11 mol m^{-3}	0.52×10^{-9}	25

across the membrane. There is, however, a basic problem in applying Fick's law to the diffusion of substances through a membrane, as the concentration gradient of the substance *in the membrane* is unknown. The closest approximation that can be obtained is the concentration gradient. between the aqueous phases on the outside (c_j^0) and the inside (c_j^i) of the membrane divided by the thickness (Δx) of the membrane (assuming a linear gradient of concentration across the membrane). Thus,

$$-\partial c/\partial x \approx -c_j^i - c_j^0/\Delta x = c_j^0 - c_j^i/\Delta x.$$

Since the concentrations at the inner and outer faces of the membrane are unknown, it is necessary to estimate these values. This is done by multiplying the concentration in the aqueous phase by a partition coefficient (K_j) which expresses the ratio of the concentrations in the aqueous phase to that in the membrane. Thus, the actual concentration gradient through the membrane is $K_j(c_j^0 - c_j^i)$. Fick's law becomes:

$$J_j = D_j K_j (c_j^0 - c_j^i)/\Delta x$$

or

$$J_j = D_j K_j/\Delta x \cdot (c_j^0 - c_j^i)$$

or

$$J_j = P_j (c_j^0 - c_j^i)$$

where

$$P_j = D_j K_j/\Delta x$$

P_j is known as the permeability coefficient. Although the values of D_j, K_j and Δx may be uncertain, P_j is a measurable quantity and may be determined empirically. Providing that the flux of a substance can be measured and the internal and external concentrations are known, P_j can be calculated. Permeability coefficients range from about $10^{-6}\,\mathrm{m\,s^{-1}}$ for ethanol to $10^{-12}\,\mathrm{m\,s^{-1}}$ for an ion such as chloride (Table 2.3). Water and many hydrated ions have permeabilities much greater than is accountable by their lipid solubility (which is close to zero) because of the presence of hydrophilic pores in the membranes: the permeability of water and such solutes does not just reflect the partition coefficient between aqueous and lipid phases.

The chief problem in estimating permeability coefficients is in obtaining the values of c_j^0 and c_j^i at the membrane surface, since it is only possible to measure these in the solution on either side of the membrane. Close to a membrane, there will be a *boundary layer* in which the concentration varies

Table 2.3 Permeability coefficients. The values for the non-electrolytes are derived from $\partial m/\partial t = PA(C-k)$, where m is the amount (mol) of a substance passing through a given area, A (m^2), in time t (s). C is the concentration external to the membrane and k is the concentration inside the cell vacuole: both are in mol m^{-3}. The values, for Characean algae, are taken from Altmann and Dittmer (1966). The values for the ions are taken from Lannoye *et al.* (1970); they are permeabilities for *Chara corallina* (= *C. australis*) in the sense that the term is used in the Goldman equation. All these values may differ widely between organisms.

Solute	Permeability coefficient ($m\,s^{-1}$)
Ethanol	$1.6–3.0 \times 10^{-6}$
Glycerol	$0.03–2.1 \times 10^{-7}$
Urea	$1.3–11 \times 10^{-7}$
Glucose	$<0.09 \times 10^{-7}$
K^+	4.8×10^{-9}
Na^+	3.5×10^{-9}
Cl^-	2.1×10^{-12}
Solvent	
Water	*c*.1×10^{-4} to $\times 10^{-5}$

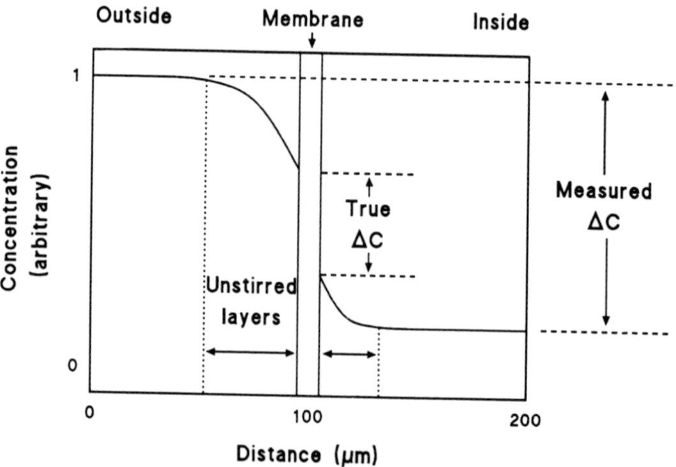

Figure 2.7 Boundary layers. The presence of 'unstirred layers' next to membranes means that the concentration of solutes at a surface cannot be measured. What can be measured is the concentration in the bulk solutions on either side of the membrane. Consequently, true gradients across the membrane cannot be calculated even if the partition coefficient for the solute between water and the membrane is known. For plant cells, unstirred layers are particularly important since their width may be amplified by the presence of the cell wall.

with distance from the membrane (Figure 2.7), so that the actual concentrations at the membrane surfaces themselves remain uncertain. The boundary layer cannot be entirely removed by rapid stirring of the solution, especially if intact plant cells are used, as the layer will be located within the cell wall, where stirring is not possible.

Fick's law describes the process of diffusion as a flux directly proportional to a driving force (the negative gradient in concentration): the flux is equal to the driving force multiplied by a coefficient (the diffusion coefficient). A similar state of affairs is seen in the passage of electricity; the current is directly proportional to a driving force, the voltage, multiplied by the conductivity (or the reciprocal of the resistance). These are examples of a general law that may be stated as 'the flux of a substance is equal to the appropriate driving force multiplied by the conductivity (or divided by the resistance)', or

$$\text{flux} = \text{conductivity} \times \text{driving force.}$$

The relationship is simple: it is only necessary to know what is the appropriate driving force.

For diffusion we used the negative gradient of the concentration, but it is easy to see that for the diffusion of an ion carrying a net charge, its movement would be influenced not only by any concentration gradient, but also by an electrical field, should one be present. For an ion the appropriate driving force would not simply be its concentration gradient, but some parameter that takes account of both concentration and charge. The appropriate driving force is a gradient of electrochemical potential. This is a part of a broader term, the *chemical potential* (μ_j), which is a measure of the free energy associated with a chemical species and available for performing work (see section 2.8.1). In any general equation relating a flux to a driving force, the appropriate gradient is therefore $-\partial\mu/\partial x$. It is a condition of equilibrium that gradients in potential do not exist, i.e. at equilibrium, $-\partial\mu/\partial x = 0$.

2.4.1 *Diffusion potentials*

Electrical forces are important because many solutes in plants carry net charge and because there is invariably an electric potential difference across membranes. If a crystal of a salt, for example sodium chloride, is added to a beaker of water, it dissolves to make a concentrated solution and the ions of sodium and chloride then diffuse through the water. The two ions diffuse at slightly different rates, however, as they have different mobilites and so a small charge separation occurs since the two ions carry positive and negative charges, respectively. This gives rise to an electrical potential, known as a diffusion potential. The effect of the diffusion potential is to accelerate the slower-moving ionic species and to retard the faster one such that electrical charge balance is preserved. Should diffusion occur across a membrane, a similar course of events will take place. One of the ions will diffuse faster than the other, setting up a charge separation and an electrical potential difference—the membrane diffusion potential. Thus electrical potentials can exist across membranes just as a result of differences in the mobility of the ions crossing the membrane. For a plant membrane, there are many ions that cross it at any given time. However, for the most part, the diffusive fluxes are dominated by movements of K^+, Na^+ and Cl^-. In this case it is possible to deduce (see, for example, Nobel, 1983) that the measured potential across the membrane (E_m) is given by:

$$E_{\mathrm{m}} = RT \ln \frac{(P_{\mathrm{K}} c_{\mathrm{K}}^{0} + P_{\mathrm{Na}} c_{\mathrm{Na}}^{0} = P_{\mathrm{Cl}} c_{\mathrm{Cl}}^{i})}{(P_{\mathrm{K}} c_{\mathrm{K}}^{i} + P_{\mathrm{Na}} c_{\mathrm{Na}}^{i} + P_{\mathrm{Cl}} c_{\mathrm{Cl}}^{0})}$$

where P is the permeability coefficient. This is the *Goldman* or *constant field equation*; constant field since it is assumed that there is a constant electric field across the membrane, viz. that $\partial E/\partial x = \Delta E/\Delta x$. It is also sometimes known as the Goldman–Hodgkin–Katz equation. The equation gives the diffusion potential across a membrane assuming independent diffusion of the ions involved down their gradients of chemical potential. If the concentrations and permeabilities of the three ions on either side of the membrane are known, it is possible to calculate whether any measured potential can be accounted for simply by the different mobilities of the ions in diffusing through the membrane. The great strength of using the equation as formulated above is that in many cases calculated and measured potentials agree rather closely. Other ions could be added in, for example Ca^{2+} or H^+, but they contribute little to the membrane potential in terms of their passive diffusion.

The diffusion potential is not the only component of the membrane potential. As we shall see, a major flux through the membranes of plant cells is of H^+, but these ions are actively transported and not simply diffusing: they do not count in the diffusion potential calculated by the Goldman equation. The fact that such a flux exists can be deduced by inhibiting active processes and demonstrating that, in the presence of the inhibitor, the membrane potential is close to the Goldman potential calculated from the fluxes of sodium, potassium and chloride (see also Figure 3.2).

2.4.2 *Nernst potential*

Cells accumulate some salts from their environment to substantial concentrations, such that the concentration inside the cell may be many times that outside the cell. When first observed, these high accumulation ratios were taken as evidence of active transport; evidence that work had to be done by the cell for the accumulation to be achieved. However, since the diffusion of ions is influenced by gradients of electrical potential as well as concentration gradients, could the existence of electrical potential gradients across the plasma membrane explain this accumulation?

Membrane potentials can be determined by inserting a glass micro-capillary connected to a suitable measuring device into a cell and recording the electrical potential differences between this capillary and the external solution. The microcapillary, or micropipette as it is otherwise known, must be small enough that it can be inserted into the cell; generally the diameter at the tip is about 1 μm where it is to be used with the cells of higher plants. The narrower the tip, however, the greater the resistance of the electrode; a 1 μm tip has a resistance of about 10^6 ohms. The capillary is filled with an electrolyte (commonly 3000 mol m^{-3} KCl rather than NaCl in order to minimise the diffusion potential at the tip). The microcapillary is really a salt-bridge that starts the connection of the plant cell to the measuring device. This connection is completed with wires and at the junction of salt-bridge and wire it is necessary to have a way of passing the current from ions to electrons. This is achieved by using a half cell, commonly silver/silver chloride. The circuit is completed by an identical half cell and a salt-bridge to the solution bathing the tissue (Figure 2.8: see Nobel, 1974 for a detailed description). The measuring device is a high input resistance (impedance) electrometer: the high impedance of the measuring device ensures that little current flows during the measurement and is necessary if an accurate measurement of the membrane potential is to be made. The potentials of the two silver/silver chloride half cells cancel.

Electrical potentials across the plasma membrane have been measured in giant algal cells. The internodal cells of certain species are several centimetres long and at least one millimetre in diameter (see Figure 2.12). In these cases it is possible to place the microelectrode in the cytoplasm with some confidence, something not necessarily possible with mature plant cells, where the cytoplasm is exceedingly narrow. The measured potentials are generally negative internally relative to the outside. This constitutes a positive driving force into the cell for a positively charged ion (e.g. K^+), but constitutes a considerable barrier for a negative ion.

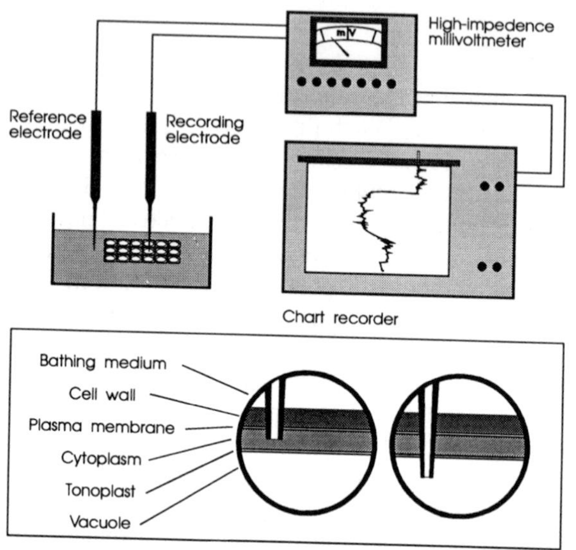

Figure 2.8 Measurement of electrical potentials across plant cell membranes. The diagram shows the arrangement of electrodes for the measurement of electrical potentials (see text for detail) and highlights the problems of inserting a microelectrode into the very thin cytoplasm present in the majority of plant cells.

It is possible to deduce to what extent a particular concentration of ions can be attributed to the existing electrical potential gradients by setting the chemical potential of the given ion equal across the membrane, a condition of equilibrium. That is, given the ion diffuses to passive equilibrium, what potential balances what concentration gradient? At equilibrium,

$$\mu^0 = \mu^i$$

and (from section 2.8.1)

$$\mu_j^* + RT\ln a_j^0 + z_i FE^0 = \mu_j^* + RT\ln a_j^i + z_j FE^i$$

from which the potential at equilibrium, E_{N_j} (named after Nernst, who first derived this relationship) is given by:

$$E_{N_j} = \frac{RT}{z_jF}\ln\frac{a_i^0}{a_j^i} = \frac{RT}{z_jF}\,2.303\log\frac{a_j^0}{a_j^i}$$

The Nernst equation illustrates the importance of the electrical component of the membrane potential. For example, at 18°C, a potential difference of just $-58\,\mathrm{mV}$ (i.e. negative inside) is poised with a concentration difference of ten-fold for a monovalent cation. Thus, since membrane potentials between the vacuole and the outside are commonly of the order of -100 mV, this would be sufficient to explain a concentration difference of about 50 times higher inside than outside.

Electrophysiological measurements can also be used to determine currents flowing across membranes. Electrodes are sited on either side of a specific membrane and a current–voltage relationship determined using a feedback amplifier (an amplifier that is able to generate a current equal in magnitude, but opposite in sign to that flowing across the membrane; see also section 2.5.1) to clamp the voltage at discrete steps, usually in the range from 0 to -300 mV. The current flowing at each voltage is plotted against the voltage. This represents the total of all electrically active pathways operating and can be manipulated, for example by changing the external conditions, to give information on the nature of the processes occurring. By comparing the current–voltage (I–V) curves before and after the addition of an inhibitor of a transport process information on the process can be obtained.

2.5 Methods of studying transport through membranes

Once membranes have been obtained, they may be used for a whole range of tests available to biochemists, from chemical analysis to electron microscopy to the assay of the activity of constituent enzymes. The application of these methods has lead, over the years, to the various models of membrane structure and to very detailed knowledge of some particular membranes, especially the thylakoids in the chloroplasts and the inner mitochondrial membranes (see sections 4.1 and 4.2; Hall *et al.*, 1982; Starzak, 1984; Yeagle, 1987). Recent developments in methodology are now providing information about the transport properties of the plasma membrane and the tonoplast. Two techniques, patch-clamping and the use of fluorescent probes, are proving particularly valuable.

2.5.1 *Patch-clamping*

Patch-clamping is a term used to describe a technique whereby a small part of a membrane, hence the term 'patch', is sealed to a hollow heat-polished glass pipette (with a diameter at the tip of about 1.0 μm; Figure 2.9). The pipette is filled with an electrolyte solution and, together with its patch, placed in a bathing solution containing a second electrode. Subsequently

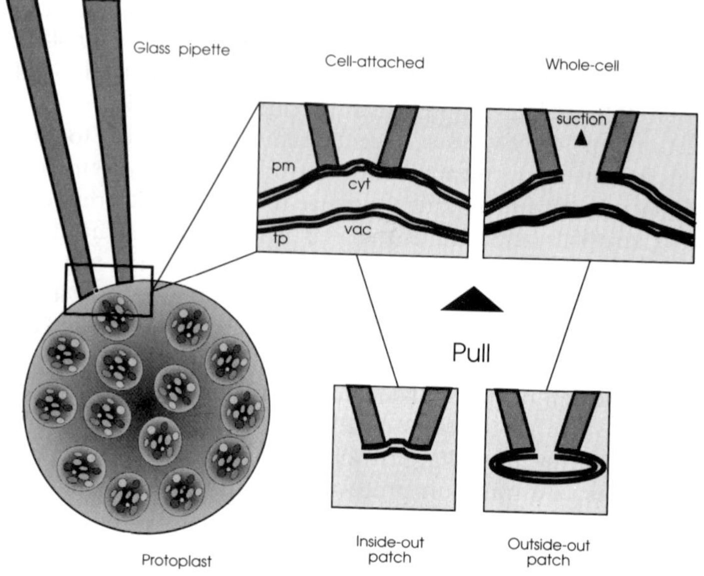

Figure 2.9 Patch-clamping. The diagram shows the attachment of a glass microcapillary to a protoplast containing chloroplasts. The detail shows the 'cell attached' and 'whole cell' configurations for the protoplast and 'patches' which are 'inside-out' and 'outside-out'.

the flow of current is measured when a series of voltages is applied between the two electrodes. Generally, the voltage across the membrane is measured using an amplifier that is then able to generate a current equal in magnitude, but opposite in sign (see Takeda *et al.*, 1985 for details). This feedback amplifier is thus able to clamp the voltage at values set by the experimenter. The technique was first used in 1976 and first applied to plant cells 1984 (see Hedrich and Schroeder,1989).

The patch is made by bringing a protoplast or vacuole into contact with the electrode and the seal made by applying slight suction to the electrode to draw the membrane into close contact with the glass. The seal, which is the consequence of interaction between the glass and the membrane at the molecular level, must have a high resistance if current flow across the membrane is to be measured rather than a leakage current. Consequently, it is often known as a *giga ohm seal* (giga as a multiplier is 10^9). The whole protoplast or vacuole may be left attached to the electrode in what is termed the 'on-cell' or 'cell-attached' configuration (Figure 2.9). Alternatively, the attachment across the end of the electrode may be ruptured so that the inside of the membrane is exposed to the solution filling the electrode: this is known as the 'whole-cell' configuration (Figure 2.9). If the electrode is pulled away from the cell-attached configuration the membrane tears, leaving a small patch across the electrode: what was anatomically the inside of the membrane is now facing the bathing solution and this is known as an 'inside-out' patch. Alternatively, if the pipette is withdrawn from the whole cell configuration, the membrane again tears, but a small part may reseal over the end of the capillary. This membrane will now present what was normally its outside face to the bathing solution and is known as an 'outside-out' patch (Figure 2.9). Where excised patches or whole-cell configurations are used, the solutions on both sides of the membrane can be regulated and information obtained about the influence of specific chemicals and their gradients across the membranes. The technique represents an extremely powerful tool in the hands of the membrane physiologist. It can be used to determine the currents flowing through single proteins, known as *channels*, in the membrane (see also section 2.6.1). Because the membrane patch is so small that it contains only one or a few proteins, the opening and closing of these individual channels can be seen: these are the rectangular pulses illustrated in Figure 2.10a. Where the current jumps between more than two levels (the two levels are equivalent to the open and closed states), this means that more than one channel is open at the same time. Computer programs are used to sort out the opening and closing from the 'noise' in any recording and to calculate the mean current carried at any given voltage. From this, the relationship between I and V (Figure 2.10b) can be plotted and used to identify the ion carrying the current. To do this, use is made of the Nernst relationship (section 2.4.2). At the point at which the net current carried is zero (where the line crosses the x-axis in Figure 2.10b, the so-called reversal potential at which are current changes direction), the influx balances efflux. By comparing the reversal potential with the calculated Nernst potentials of the different ions present in the system (from the known activities either side of the patch) the ion carrying the current can be identified. The duration of opening and closing is affected by specific chemicals and by the voltage difference across the membrane.

Although channels can be opened and closed at will under experimental conditions, we have limited knowledge of the factors which cause these changes *in vivo*. In many cases the voltages producing channel activity in a patch are quite unphysiological, whereas at expected physiological voltages (membrane potentials) the channels have a small probability of being open (see Satter and Moran, 1988). Such anomalies probably reflect losses of important regulatory mechanisms during the patching procedure.

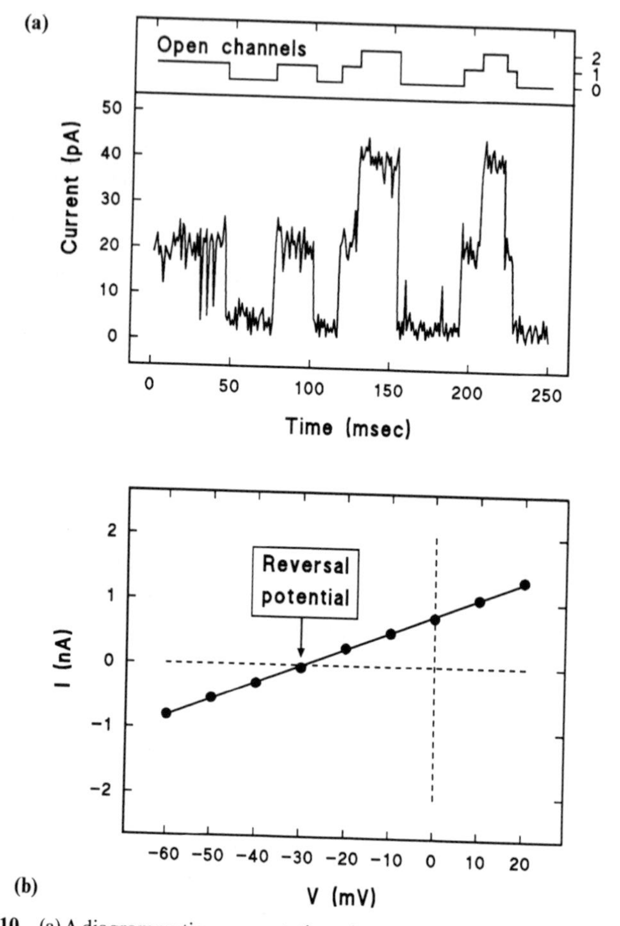

Figure 2.10 (a) A diagrammatic representation of recordings from inward rectifying channels of vacuoles of *Vigna unguiculata* in the 'cell-attached' configuration. (b) The relationship between the current (I) and the voltage (V) for vacuoles of *Plantago media* in the 'cell-attached' configuration showing the potential at which the net current is zero—the reversal potential. Drawn from the data of Maathuius (1991).

2.5.2 *Fluorescent probes*

Patch-clamping requires the preparation of isolated protoplasts or vacuoles and cannot give information about the transport of materials across the tiny vesicles such as those prepared by cellular disintegration and density gradient centrifugation (the membrane and micropipette have to be visible with a light microscope for the operator to bring them into contact). Information can be obtained from vesicular systems, however, by the use of chemicals that change their properties in relation to their local environment. The most convenient property to monitor is fluorescence: the chemicals are known as fluorescent probes. Fluorescence is the emission of photons with a characteristic spectrum following excitation of the molecule by photons with a higher, and again characteristic, energy spectrum. The bulk of the work so far carried out with plant membrane systems has been concerned with monitoring changes in pH associated with gradients.

The first probes to be used were amines, which are highly permeant in their non-ionic form but become much less permeant if they are protonated. They pass through a membrane and enter a vesicle in the non-ionic form. They will accumulate in the vesicle if its inside is acid or becomes acid because the chemical is protonated in these conditions and becomes 'trapped' inside the vesicle. Changes in fluoresence occur due to changes in the concentration of the dye within the vesicle and in some cases due to the effects of pH itself on the probe.

The application of fluorescent probes is a rapidly developing field with the synthesis of new and better fluorochromes (which may be quite complex organic molecules; Figure 2.11) and advances in optics and electronics. Both fluorescence and excitation spectra may change as the result of subtle interactions of the probe with ions, such as protons and calcium, in the cell. The changes in the excitation spectrum upon binding of the probe are the basis of fluorescence ratio imaging. Figure 2.11 shows the spectra of the calcium probe fura-2. As calcuim binds to the probe, the excitation spectrum shifts to shorter wavelengths but the emission spectrum is not affected. The degree to which the probe is saturated with calcium can be deduced from the ratio of the amount of fluorescence detected upon alternating illumination with a longer and a shorter wavelength. This characteristic has an enormous advantage in quantification because information about concentration is obtained independently from the total emission intensity. Theoretically the ratio is independent of the concentration of the probe and the size of the object studied. By scanning the fluorescence image with a television camera and analysing the signal digitally, concentration maps can be produced in pseudo-colour and displayed continuously.

The use of fluorescent probes has to be viewed with some caution as there are many opportunities for artefacts to arise from non-specific binding with other chemicals or cell constituents. For example, in the determination of pH, fluorescence may change with the change in environmental parameters

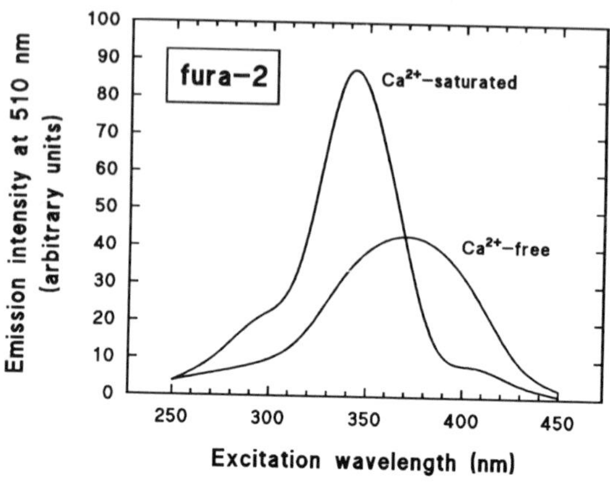

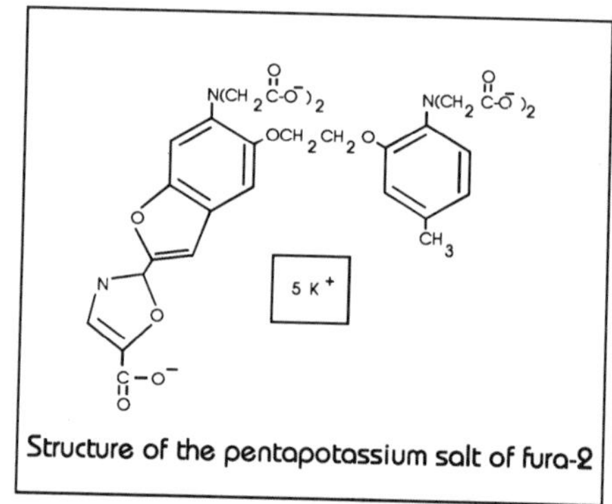

Figure 2.11 Structure and spectra of the Ca-sensitive fluorescent probe fura-2. Redrawn with permission from Haugland (1989).

other than just pH. For example, the binding of the dye to the membrane may be altered by the concentration of mono- or divalent cations.

2.6 Properties of plant membranes

Although the appearance in the microscope of membranes from plant cells is well described, their composition, both chemical and physical, is less well defined than for animal cells. This is partly due to the problems mentioned earlier of preparing samples of pure membranes for analysis. Plant cells do, however, have a broadly similar membrane composition to animal cells,

Table 2.4 Composition of some plant membranes (%molar basis).[a]

Membrane and origin	Sterols[b]	Phospholipid	Galactolipids[c]
Barley[d]			
Root plasma membrane	57.2	17.5	16.8
Leaf plasma membrane	34.8	39.6	21.4
Barley root[e]			
Plasma membrane	31.2	43.3	25.4
Tonoplast	10.7	61.9	27.5
Endoplasmic reticulum	3.7	78.7	17.6
Spinach leaf[d]			
Plasma membrane	7.0	59.0	29.2
Mung bean hypocotyl[f]			
Plasma membrane	43.6	48.9	7.6
Tonoplast	27.8	51.1	21.1

[a] other lipids take the total to 100%
[b] includes sterol esters, acylated sterol glycosides and sterol glycosides
[c] mono- and di-galactosyl diglyceride
[d] Rochester *et al.* (1987)
[e] Brown and DuPont (1989)
[f] Yoshida and Uemara (1986)

with the same classes of lipids (Table 2.4) and a range of proteins. Typically, the plasma membrane is the thickest of the plant cell membranes, often with a width of 10 nm. The other membranes range in thickness from about 6 to 7.5 nm (Table 2.1). The two fracture faces of the tonoplast are different, the cytoplasmic face having a higher number of intramembranous particles per unit area than the luminal face.

The early work on membrane transport in plants was helped by the analysis of the movement of substances into and out of the giant cells of certain algae such as *Nitella* (Figure 2.12) and *Chara*. From these early studies it was clear that the membranes were highly permeable to water, whereas their permeability to non-electrolytes was generally directly related to their lipid solubility. Electrolytes normally permeated less readily than non-electrolytes, although in many cases much more rapidly than could be accounted for by their lipid solubility. It also became clear that plants in general, not just algae, had a degree of control over their content of electrolytes. For example, as the concentration of potassium in the external solution around the roots of a variety of higher plants (14 species) is changed from 8 to 1000 μM, the concentration in the plant tops only changes from 109 ± 8.9 to $190 \pm 8.6 \mu mol\, g^{-1}$ fresh weight (Asher and Ozanne, 1967; Figure 2.13).

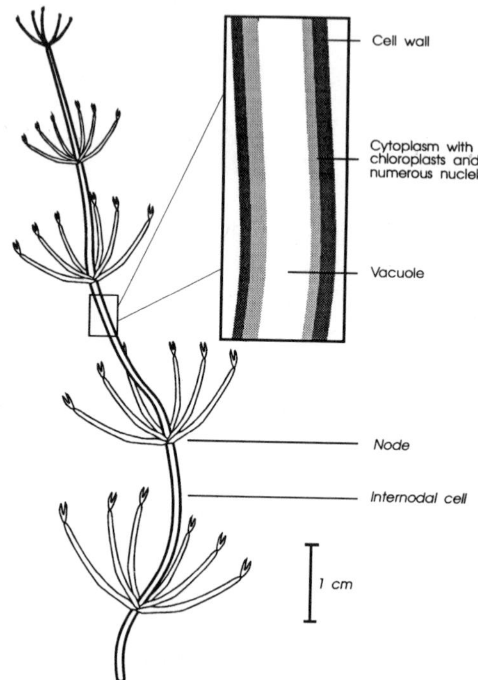

Figure 2.12 A drawing of a vegetative plant of a species of *Nitella*, showing the large internodal cells. Species in the genus *Chara* appear similar, but with strands of smaller cells surrounding the giant internodal cells.

2.6.1 *Ion channels, carriers and pumps*

Ions cross membranes much more freely than they might be expected to diffuse through the lipo-protein matrix and do so with considerable selectivity and specificity. The current view is that most are conducted across the membrane through passageways composed of protein and termed *channels*. Movement may simply be due to an electrochemical gradient or it may require energy for the transport to occur. In cases where metabolic energy is expended *directly* in the transport of a substance across the membrane, we will describe the process as involving a *pump*. An ion channel may be an integral part of a pump and this is potentially a cause of some confusion (a proton, for example may be pumped through a channel; cf. section 3.1). Nevertheless, the involvement of the metabolic energy in a channel defines that particular channel as part of a pump. Ionic channels

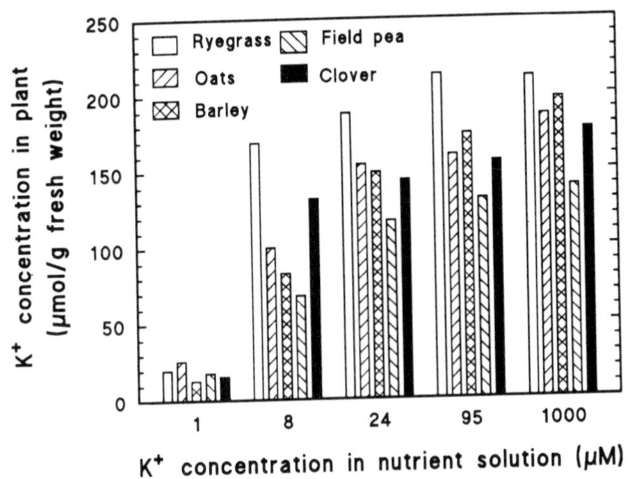

Figure 2.13 The effect of changes in the external concentration of potassium on the potassium concentration in the tops of five plant species. Large changes in the external concentration (three orders of magnitude) have rather little effect on the internal concentration. Drawn from the data of Asher and Ozanne (1967).

are intrinsic proteins that continuously change conformational state. When the protein, which spans the membrane, forms a pathway though which the ion can diffuse, the channel is said to be open. Channels are only open for a short period in the life of an organism, but the probability of opening and closing (*gating*) can be affected by a change in potential across the membrane (voltage-dependent or voltage-gated channels), by the binding of a specific chemical (receptor-operated channels), by physical factors (stretch-operated channels) and, we presume, by factors of which we are not yet aware. Some chemicals activate channel opening (*agonists*) and some inhibit (*antagonists*). Although the opening and closing can be under metabolic control, the movement of the ion is simply down a gradient of free energy: the driving force is already present and the change in conformation simply opens or closes a gateway through which ions can traverse the membrane barrier. Ion movement through channels that are not part of pumps is at the rate of about 10^6 to 10^7 ions per second. Pumps, on the other hand move solutes at rates of 10^4 to 10^5 transported molecules or ions per second: they transfer just one or a few ions or molecules per catalytic cycle.

There is yet another class of membrane proteins involved in the movement of solutes across membranes: these are known as *carriers*—

membrane proteins that bind, covalently, to the substance being transported. Carriers function at similar rates to pumps and are also known as *translocators*, *permeases*, and *porters*. Porters may be *symporters*, where two ions or molecules cross the membrane in the same direction or *antiporters* where transport of two substances is in opposite directions. In both cases the link between the movement of the two transported substances is obligatory. A carrier, like a channel may or may not be part of a pump.

All these transport processes involving integral membrane proteins are sometimes known as *mediated* transport processes: transport is mediated by the presence of the protein in the membrane. However, some substances can simply diffuse through the membrane; they just have to be soluble in the lipid phase. Examples of such lipid soluble substances are some of the fluorescent probes mentioned earlier and used in connection with monitoring changes in pH and Ca^{2+} across membrane vesicles (section 2.5.2). Another interesting example is ammonia. Although ammonia exists in solution as the ammonium cation, it is in equilibrium with free ammonia. The latter is able to diffuse across membranes and once across it re-establishes an equilibrium with its ammonium ion.

outside membrane inside

$$NH_4^+ \longrightarrow \underset{\downarrow}{H^+} + NH_3 \xrightarrow{\quad \| \quad} NH_3 + \underset{\uparrow}{H^+} \longrightarrow NH_4^+$$

The net effect is the creation of a proton gradient across the membrane. Most ions are not lipophilic, however, and cannot simply diffuse across biological membranes.

2.7 Water and osmosis

Although the discussion so far has centred on a description of the movement of solutes, their fluxes cannot be divorced from fluxes of water. Biological membranes are rather permeable to water relative to the common solutes such as sucrose, glucose, sodium, potassium and chloride (Table 2.3). This differential permeability is often described as *semi-permeability*: the membrane is more permeable to solvent than solute. Semi-permeability is the basis of osmosis. The movement of solute across a membrane changes the free energy of the water on one side of the membrane, relative to the other side. During osmosis, water movement is brought about because water moves through the membrane down a gradient of free energy. For water the free energy gradient driving water movement is normally described, for historical reasons, in pressure units and is known as the water potential, ψ_w (section 2.8.2).

The molecular mechanisms by which water crosses biological membranes have remained a matter of debate for many years, although the thermodynamic description of osmosis is well developed. Water must pass through pores that exclude solute. The pores therefore contain pure water and are in contact with solution at their internal and external mouths, although their structure in plant membranes is not physically defined. Thermodynamics clearly show that a concentration difference is balanced by a pressure difference and it is this that drives water through the membrane. At the molecular level, water is pulled through the pore by attraction to a solute molecule whenever a solute molecule is at the mouth of the pore (Dainty and Ferrier, 1989).

The movement of water across membranes is an important aspect of transport processes occurring in plant cells. In this chapter we have concentrated on the nature of membranes, how they can be obtained and studied and the forces that drive transport. In the next chapter, we will examine the properties of some plant membranes in much more detail.

2.8 Addenda

2.8.1 *Chemical potential*

In any system, be it plant, plant cell, or the soil-plant-atmosphere continuum, work is associated with change. As systems change from one state to another they undergo a change in their free energy. Under conditions of constant pressure and temperature—to which most biological systems approximate, at least in the short term—an appropriate measure is the Gibbs free energy. This always *decreases for a spontaneous process* at constant temperature and pressure. Unfortunately, the Gibbs free energy is an extensive property, i.e. like mass, the more of a substance there is the more its Gibbs free energy. A more useful quantity is the *chemical potential*, i.e. the rate of change of Gibbs free energy with the number of moles of a substance when factors that influence the free energy, such as temperature (T), pressure (P), electrical potential (E), height in a gravitational field (h) and amounts of other substances (n_i), are held constant. Put mathematically:

$$\mu_j = \frac{(\partial G)}{(\partial n_j)_{P,T,E,h,n_i}} \qquad (1)$$

from which, in solution,

$$\mu_j = \mu_j^* + RT \ln a_j + \bar{V}_j P + z_j FE + m_j gh \tag{2}$$

where μ_j^* is an (unknown) constant (from integration) and is actually equivalent to the chemical potential of the substance in a standard state, a_j is the activity, $\bar{V}_i$ is the partial molal volume of the substance and analogous to the molecular weight, z_j is the valency, m_j is the molecular weight of the substance, g is the acceleration due to gravity and R is the gas constant.

2.8.2 *Water potential*

The water potential terminology devised by Slatyer and Taylor in the early sixties was based upon the principles of thermodynamics. From the equations in section 2.8.1, it can be shown that

$$\mu_w = \mu_w^* - V_w \pi + \bar{V}_w P + m_w gh \tag{3}$$

where μ_w^* is an (unknown) constant (from integration) and is actually equivalent to the chemical potential of water in a standard state (the activity of water is 1, the pressure is atmospheric, the electrical potential is zero, the height in the gravitational field is zero and the temperature is that of the system under consideration), $\bar{V}_w$ is the partial molal volume of water (about 18 cm^3 per mol and analogous to the molecular weight), m_w is the molecular weight of water and the other symbols are as defined in section 2.8.1.

Water then moves down gradients of free energy from high to low and it is a condition of equilibrium that gradients of free energy are zero.

In order to make the concept easier to use, water potential was devised as the difference in free energy of pure free water and that of water in the system under consideration. Thus

$$\psi \mu = \frac{\mu_w - \mu_w^*}{\bar{V}_w} \tag{4}$$

When equation (3) is substituted in (4), the relationship between water potential, osmotic pressure and turgor pressure is derived, plus a term that is related to the effect of gravity:

$$\psi_w = P - \pi + \rho_w gh \tag{5}$$

where ρ is $m_w/\bar{V}_w$ or the density.
The equation is frequently written as

$$\psi_w = P + \pi + G \tag{6}$$

where π is the osmotic potential (the negative of the osmotic pressure) and G is the gravitational potential ($\rho_w gh$). It is easy to show that the gravitational potential is only 0.01 MPa per metre change in height (see below), so this component of the equation can often be ignored, reducing the equation to:

$$\psi_w = P + \pi$$

The gravitational potential, G, is given by

$$G = m_w gh,$$

where m_w is mass per mole, 18.016 g mol^{-1}, g is the acceleration due to gravity, 9.8 m s^{-2} and h is the vertical height in m.

Thus, per metre of vertical height, the gravitational potential changes by $0.018016 \times 9.8 \times 1$ (kg mol^{-1}) (m s^{-2}) (m) or kg m^2 s^{-2} mol^{-1} or J mol^{-1}. To convert this to pressure units, divide by the partial molal volume, $\bar{V}_w$, which is 1.805×10^{-5} m^3 mol^{-1} at 20°C, i.e.

$$(1.8016 \times 9.8 \times 10^{-2})/(1.805 \times 10^{-5}) = 9.8 \times 10^3 (\text{J mol}^{-1})(\text{m}^{-3}\text{ mol}) \text{ or J m}^{-3}.$$

Since a Joule is a N m, a J m^{-3} is the same as a N m^{-2} or a Pa. Thus the value of G changes by 9.8 kPa per metre or approximately 0.01 MPa per metre.

The greatest difficulty generally encountered in using the water potential concept is a semantic one. Since the water potential of pure free water is set to zero, the water potential of most plant cells is negative. This causes semantic problems concerned with higher and lower. It is useful to compare water potential to sub-zero temperatures: $-20°$ is clearly *lower* than $-10°$. For most of us it is then easy to judge what is higher and lower. So, a water potential of -1.0 MPa is lower than one of -0.5 MPa.

The concept of water potential has been particularly useful in descriptions of water movement from soil through the plant to the atmosphere. Much of this has undergone little change over the years.

CHAPTER THREE

ION TRANSPORT ACROSS THE PLASMA MEMBRANE AND TONOPLAST

The solutes accumulated by plant cells may be in the cytoplasm, including the subcellular organelles such as the chloroplasts and mitochondria, but the majority are in the vacuoles, as the vacuoles are the largest compartments in mature plant cells. To be accumulated, the vacuolar solutes must have crossed both plasma membrane and tonoplast. The properties of plasma membrane and tonoplast are, therefore, crucial to the solute relations of plant cells.

3.1 Active transport

Solutes may cross plasma membrane or tonoplast through a channel or via a carrier: they may diffuse or be pumped (see section 2.6.1). If they pass through a channel, their transport is down a free energy gradient. The movement is *passive*, by diffusion, although the free energy gradient may have been created by other processes such as those contributing to the membrane potential. If work is done to move the ion from a region of lower to higher chemical potential, then the transport process is said to be *active*. There are two levels of active transport. *Primary active transport* involves the direct use of metabolic energy to pump the solute across the membrane. In *secondary active transport*, the free energy gradient generated by the primary transport process is harnessed to move other solutes through the membrane via the carrier or channel proteins. It is worth emphasising the difference between *primary* and *secondary* active transport by an example. In Figure 3.1 the pumping of protons across the membrane consumes adenosine triphosphate (ATP) and generates a difference in electrical potential across that membrane. Because of this membrane potential, potassium can enter through an ion channel down its free energy gradient. The proton is pumped, a primary active process, and movement of the potassium is dependent on primary active transport. This results in

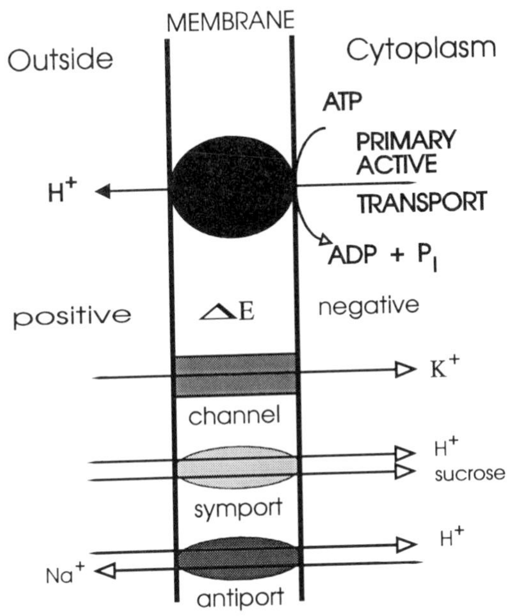

Figure 3.1 Primary and secondary active transport. The diagram depicts a primary electrogenic active transport of protons across a membrane. Energy in the form of ATP is used and the transport generates a membrane potential, ΔE. The proton gradient may be utilised to drive other transport events. In the example, uptake is through a uniport (K^+) or symport (sucrose) or efflux via an antiport (Na^+).

secondary active transport and would not have happened if the protons had not been pumped, although the transport of potassium is itself passive. It has become apparent that cells have relatively few primary transport systems and that the nature of the primary system is different in different classes of organism (bacteria, animals and plants).

A key question, then, is which solutes are transported actively across plant cell membranes? Evidence accumulating since the 1970s has favoured the movement of protons as the primary active process. Proton transport uses ATP and is electrogenic; it generates a membrane potential. This membrane potential may be sufficient to alter the electrochemical potential of an ion such as K^+ so that given the presence of an open channel, it may diffuse into the cell. However, the proton pump also generates an electrochemical potential gradient of protons. The energy available when protons return across the membrane down their gradient in free energy may be coupled to the movement of another solute against its gradient in free energy. The solute (anion, cation or non-electrolyte) may move in the same direction as the protons, in which case the carrier is a *symporter*, or in the opposite direction, in which case the carrier is an *antiporter* (Figure 3.1). If, for each proton, a single negative charge moves in the same direction or a single positive charge moves in the opposing direction, then the carrier is electrically neutral. In all other cases the carrier is electrogenic. If the movement of a proton, for example, always occurred with an accompanying hydroxyl ion or bicarbonate ion, transport would be electrically neutral. For electrogenicity, there must be the net movement of charge across the membrane.

How is it possible to determine whether a measured membrane potential has an electrogenic component? There are a number of tests that can be made. The two which are most direct are (i) that the membrane potential should be considerably more negative than the diffusion potential calculated from the Goldman equation (see section 2.4.1) and (ii) that the measured membrane potential should be reversibly depolarised (made less negative) by metabolic inhibitors (chemicals that will reversibly interfere with the energisation of any electrogenic pump) (Figure 3.2; see Lüttge and Higinbotham, 1979). Using such criteria there is good evidence (Spanswick, 1981) that electrogenic pumps are common in algae, fungi and higher plants.

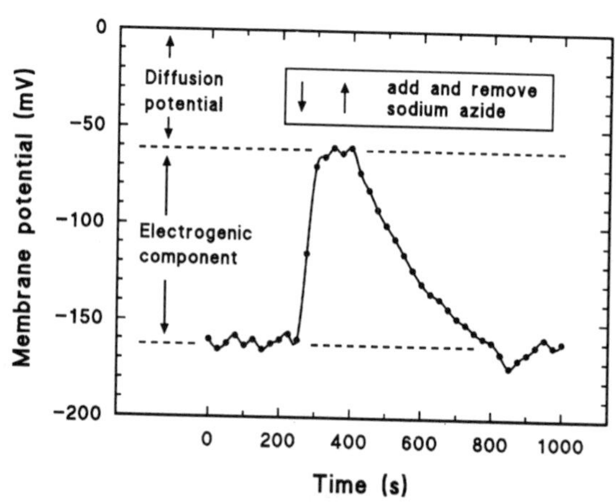

Figure 3.2 A schematic diagram illustrating the evidence for the existence of electrogenic pumps in plants. The addition of a metabolic inhibitor, in this case azide, depolarises the membrane potential to about the value of the diffusion (Goldman) potential. Subsequent removal of the poison results in the re-establishment of the pump and the hyperpolarisation of the membrane potential.

In principle, it is possible to identify a specific electrogenic pump if the ion that is transported can be removed from the pump, as current flow across the membrane will then decrease. An electrogenic inwardly directed chloride pump in the marine alga *Acetabularia* has been identified by this means (see section 3.4.2). Although this is a satisfactory procedure for a pump that works from the outside of the plasma membrane, it is virtually impossible to manipulate the ions within the cytoplasm so that those involved in outwardly directed pumps (to the cell wall or the vacuole) can be identified. Perfusion techniques have been attempted in the giant-celled algae such as *Chara* and *Nitella*, but in higher plants the manipulation of these other pumps can be achieved only by using isolated intact vacuoles or vesicles of membrane. The quest for the nature of outwardly directed electrogenic pumps is therefore difficult and, in plants, has largely been a pursuit of the last decade.

By the early 1960s it was clear that ion pumps in animal cells involved ATPases (enzymes that hydrolysed ATP during the operation of the pump) driving, for example, the transport of sodium. Since electrophysiological studies on plants have generally followed the lead from studies on animals there was speculation at that time that similar ATPases were to be found in plants. By the mid-1970s, however, there was evidence that the efflux of protons coupled to the hydrolysis of ATP was a key process in transport in bacteria and fungi. This and the general acceptance of the chemi-osmotic hypothesis (Mitchell, 1966) with its central role for proton transporting ATPases led to a re-evaluation of the importance of electrogenic proton pumps in plants (Poole, 1978; Hanson, 1978). Around the same time, ATP-dependent electrogenic proton pumping at the plasma membrane of the alga *Chara* and of the fungus *Neurospora* was demonstrated. So by the 1980s it was apparent that the primary active transport in plants and fungi was more akin to that in bacteria than that in animals and that proton transport had a central function.

3.2 Proton pumps

The net movement of protons is easy to monitor in some respects because pH (the concentration of protons) is readily measured with electrodes (and a pH meter) and there are fluorescent probes that react to changes in their environmental pH (see section 2.5.2). It is very difficult, however, to determine unidirectional fluxes of protons in plants. This is necessary if the protons moved by the pump itself are to be separated from the simultaneous flux of protons back into the cell down their electrochemical potential gradient, possibly via symporters. Unidirectional fluxes can be estimated in animal cells from electrophysiological measurements, but it generally has not been possible to use the same methods with plants, because of the difficulty of placing electrodes in the cytoplasm of plant cells (see Figure 2.8). The radioisotope methods normally used to determine unidirectional fluxes of other ions (cf. section 1.2.2) cannot be used, as the only possible tracer is tritium. This must be supplied as tritiated water (^3H_2O) to which the membranes have a very high permeability, as they do to water. Further, there are passive proton fluxes and movement of OH^- and HCO_3^- from fluxes of CO_2. Consequently, as only net movement of ^3H can be measured, movement of protons through any electrogenic pump is indistinguishable from all these other fluxes. These factors have all slowed progress to understanding primary active transport in plants. Recent advances have been aided by the use of isolated membranes. Fluorescence imaging techniques (see section 1.2.2) offer good prospects for the future.

3.2.1 *Use of membrane vesicles*

By using isolated vesicles of plasma membrane and tonoplast (see section 2.3), not only can the properties of the enzymes present in the membrane be investigated, but their transport capabilities can also be examined. There are, however, important limitations to carrying out transport studies on isolated vesicles (Sze, 1985), especially those produced from density gradients of cell homogenates. Firstly, the origin of the vesicles is generally equivocal; tonoplast and plasma membrane are usually mixed to varying degrees with each other and with other membranes. Another concern is the orientation of the vesicle, whether what was *in vivo* the cytoplasmic face of the membrane is that presented to the medium, the outside of the vesicle *in vitro*. This matters because the pumps are directional, pumping into or out of the cytoplasm both at plasma membrane and tonoplast. A further uncertainty is whether the permeability of the isolated vesicle matches that *in vivo* or whether it is more permeable than the native membrane from which it originates. Membrane constituents may be lost during isolation and it is not possible to predict the consequences of removing the cytoplasm on the membrane, or of metabolic regulation by the cytoplasm.

In spite of all these difficulties, however, it has been clearly demonstrated that vesicles enriched in both plasma membrane and tonoplast will transport protons in the presence of ATP and Mg^{2+}. There are a number of lines of evidence for this: the first is indirect. ATPase activity associated with membrane vesicles is enhanced in the presence of molecules that facilitate the transport of protons across the membrane (*protonophores*, a class of *ionophores*); in effect these molecules make the membrane leaky to protons. This observation is consistent with the following model: the ATPase is sited such that it reacts with ATP on the outside of the vesicle—the natural cytoplasmic or c-side of the membrane—and protons are transported into the vesicle, such that the concentration of protons inside the vesicle builds up. Continued transport is more and more difficult, as the free energy gradient against which transport occurs increases. If the high internal electrochemical potential of protons is dissipated by the loss of protons from the vesicle mediated by the protonophore, then the constraint is removed and ATPase activity will be enhanced.

More direct evidence comes from following the build-up of differences in electrical potential and of the concentration of protons themselves across the vesicle membrane. Electrical potentials cannot be measured across small membrane vesicles using electrodes, so indirect methods must be used to detect the potential differences. This has been achieved with two different approaches. Firstly, if the vesicles are mixed with an anion or cation that is able passively to permeate the membrane, the ion will distribute according to the Nernst equation (section 2.4.2):

$$E_{\mathrm{Nj}} = \frac{RT}{z_j F} \ln \frac{c_j^0}{c_j^i}$$

where c_j^0 is the concentration in the medium and c_j^i the concentration inside the vesicle. For a given external concentration of the permeant ion, the concentration inside will be governed by the membrane potential. This permeant ion is a simple molecular probe. The experimental protocol is to mix vesicles, probe, ATP, Mg^{2+} and K^+ and subsequently to separate vesicles from solution. The amount of the probe associated with a given volume of vesicles is measured, usually by measuring the amount of radioactivity after using a radio-labelled probe and calculating its internal concentration. Thiocyanate (an anion) is commonly used where the vesicle is inside-positive and triphenylmethylphosphonium (a cation) to detect inside negative vesicles. Allowance must be made for the amount of the probe in spaces around the vesicles (not always as easy as it sounds) and care taken that the probe does not simply bind irreversibly to the membranes (it should be possible to wash out all of the probe). Such experiments have shown that ATP generates membrane potentials. These experiments have been corroborated by the second approach; measurements obtained using fluorescent probes.

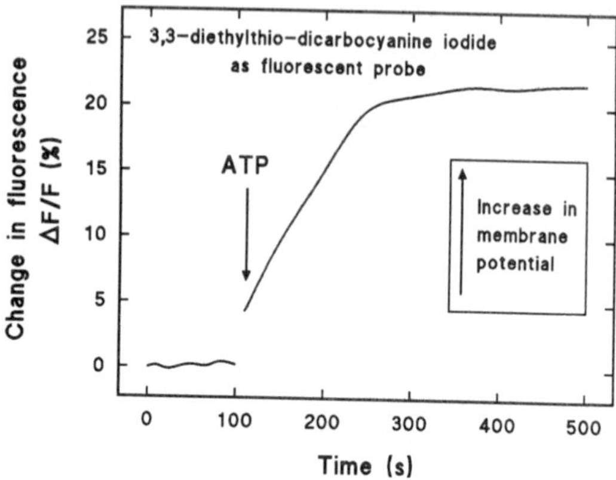

Figure 3.3 Evidence for the generation of an electrical potential difference across the tonoplast of *Beta vulgaris* (red beet) on addition of ATP. Whole vacuoles were incubated with a cyanine dye and the change in fluorescence recorded. Addition of ATP brought about an increase in the membrane potential from about −20 mV to about −13 mV. Redrawn from the data of John and Miller (1986) with permission.

The advantage of following changes in membrane potential using fluorescence rather than determining the distribution of lipophilic ions is the speed of response; separating vesicles from solution takes a minimum of ten seconds, whereas fluorescence changes can be followed well within a second of changing the external medium. The quenching of the fluorescence of cyanine dyes and of derivatives of oxonol has been used to follow changes in the membrane potential in vesicles enriched in plasma membrane and tonoplast and in whole vacuoles of red beet on addition of ATP. These measurements show changes in fluorescence on adding ATP (Figure 3.3) and support the conclusion that there is an electrogenic pump present both in plasma membrane and tonoplast (Giannini and Briskin, 1987).

If the ATPase is a proton pump, then its activity should generate not only a gradient of electrical potential, but also a proton concentration gradient. As in the case of demonstrating the presence of ATP-generated membrane potentials, techniques using both radio-labelled and fluorescent probes have been valuable tools in revealing ATP-dependent movement of protons across membrane vesicles. Just as permeant ions can be used to determine the membrane potential, weak bases can be used to determine vesicles which are interior-acid and weak acids to determine interior-alkaline vesicles. The probes distribute according to the Henderson–Hasselbalch equation:

$$pH_{in} = pH_{out} - \log \frac{[\text{probe}]_{in}}{[\text{probe}]_{out}}$$

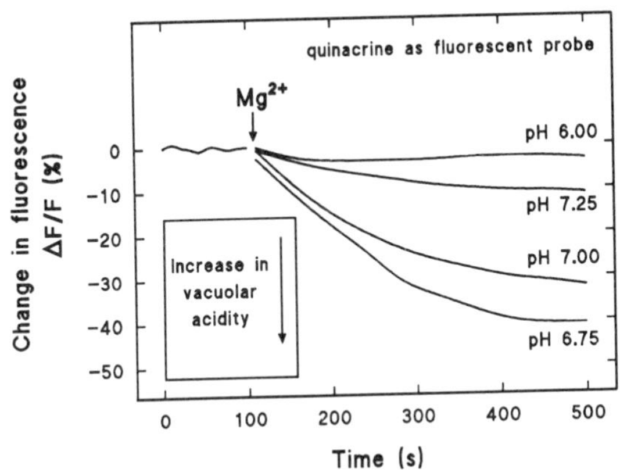

Figure 3.4 Evidence for the generation of a proton gradient on addition of ATP to vesicles of membrane prepared from the halophyte *Suaeda maritima*. A membrane fraction enriched in vesicles of tonoplast was incubated with the dye quinacrine in the presence of ATP and changes in fluorescence recorded on adding Mg^{2+} (the substrate of the ATPase is Mg-ATP, so a convenient way to start the reaction is by the addition of magnesium ions). Proton-pumping into the vesicles was determined at various pH values and was maximal at pH 6.75, somewhat lower than for most tonoplast V-type ATPases. Data of Dr W.J. Rogers University of Sussex.

The concentration of the probe inside ($[\text{probe}]_{in}$) and outside ($[\text{probe}]_{out}$) the vesicles is used to estimate the internal pH. As in the case of the measurement of the membrane potential, optical probes offer a much more rapid response to the movement of protons. Two compounds, acridine orange and quinacrine, have been widely used to demonstrate the ATP driven movement of protons across membrane vesicles (Figure 3.4).

At present the analysis of solute fluxes across the membrane of small vesicles is limited by quantification: it is impossible to compute the flux of protons per unit of membrane surface area. Measurements are made on aliquots of vesicles for which, at best, there may be a knowledge of the quantity of membrane protein. The situation is worsened where, as is common, this is a population of vesicles of uncertain origin, sidedness and leakiness.

3.2.2 *ATPase*

Cytochemical studies with the electron microscope have shown ATPase activity to be associated with membranes from plant cells. The procedure works by supplying the substrate (ATP) which is hydrolysed by the ATPase to release phosphate. The phosphate is then reacted with a soluble lead salt to produce insoluble lead phosphate at the site of the reaction. This lead

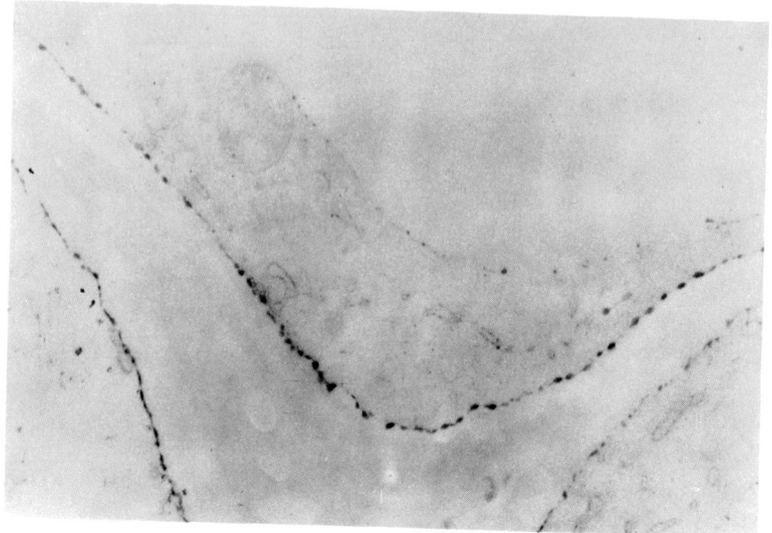

Figure 3.5 Cytochemical staining for ATPase activity. Sections of pea root were incubated with ATP and then stained with lead nitrate. Dense deposits of lead phosphate can be seen associated particularly with the plasma membrane. Magnification: × 20 000. Electron micrograph kindly provided by Dr J. Thorpe University of Sussex.

phosphate is electron dense and so visible in an otherwise unstained section as a dark deposit (Figure 3.5). However, it is not easy to know whether the enzyme whose activity has been detected is an ATPase associated with a transport process or some other ATPase, such as a non-specific phosphatase. Again, progress has waited on the characterisation of membrane fractions.

The use of isolated membrane vesicles has shown three clearly different types of ATPase to be associated with specific membrane fractions isolated from algae and the roots, leaves and cultured cells of a variety of higher plants (Table 3.1; Sze, 1985; Taiz *et al.*, 1989). One type of ATPase has long been known through its presence in organelles such as mitochondria and chloroplasts. This is the so-called *F_0F_1 type ATPase* (or *F-type*) that couples proton pumping to ATP synthesis. More recently discovered, in plants at least, are two other types of ATPase, the *P-type ATPases* and the *V-type ATPases*, differing in their reaction to vanadate (at a concentration of about 5 μM): one type inhibited and the other is not. Both types of enzyme are insensitive to molybdate, an inhibitor of non-specific phosphatases and to oligomycin, an inhibitor of the F_1 component of the $F_0\,F_1$-

Table 3.1 Some properties of proton translocating ATPases found in plant cells.

Type site	P plasma membrane	F_0F_1 mitochondria	V tonoplast
Normal function	H^+pumping	ATP synthesis	H^+pumping
Molecular mass (kDa)	100	500	600
Subunits	1–2	10–12	3–9
pH optimum	6.5	8.0	7.5
Inhibitors			
molybdate	no	no	no
vanadate	yes	no	no
oligomycin	no	yes	no
DCCD	no	yes	yes
Phosphorylated intermediates	yes	no	no
Effect of			
K^+	+[a]	=	=
Cl^-	=	+	+
NO_3^-	=	−	−

[a] + sign is used to indicate stimulation; − to denote inhibition; = sign indicates the ion has little effect.

type ATPase found in mitochondrial membranes. The vanadate-sensitive ATPase is characteristic of the plasma membrane (P-type), while the vanadate-insensitive ATPase is a feature of the vacuolar membrane (hence, V-type). Cytochemical studies tend to confirm the localisation of the vanadate-insensitive enzyme at the tonoplast and the vanadate-sensitive enzyme at the plasma membrane, although much activity is lost due to the fixative glutaraldehyde used to prepare the material for electron microscopy (Balsamo and Uribe, 1988). Both types of enzyme are capable of bringing about the transport of protons across vesicle membranes.

3.2.2.1 *P-type ATPases.* The plasma membrane ATPases are members of a class of ATPases known as P-type ATPases (or P-ATPases or E_1E_2 ATPases or (E–P) ATPases). They are sensitive to vanadate as this anion interferes with the formation of a phosphorylated intermediate necessary to the function of the enzyme. Activity is optimal around pH 6.5, stimulated by cations ($K^+ > NH_4^+ > Rb^+ > Na^+$) and the fungal phytotoxin fusicoccin, but not by anions. Activity requires Mg^{2+}, with Mg-ATP being the substrate for the enzyme (see Briskin, 1990). The plasma membrane proton-translocating ATPases (designated EC 3.6.1.35 by the Enzyme Nomenclature Committee of the International Union of Biochemistry) consist of a

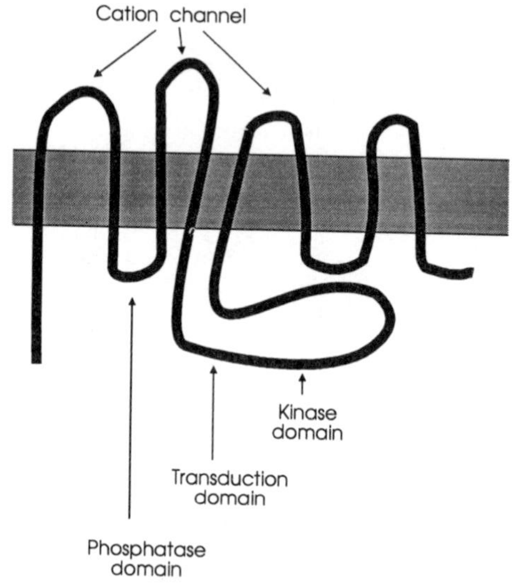

Figure 3.6 The putative arrangement of the proton-translocating ATPase in the plasma membrane, derived from hydropathy plots of the amino acid sequence of the enzyme from *Arabidopsis thaliana.*

single catalytic subunit of about 100 kDa, although how many subunits exist together in the native state remains uncertain (Briskin, 1990). The protein, which has been cloned and sequenced, is thought to cross the membrane eight times with both N- and C-termini lying in the cytoplasm (Figure 3.6; see Sussman and Harper, 1989). Following the sequencing it has been possible to compare the plant proton plasma membrane ATPase with other E_1E_2 ATPases. Although overall homologies are not great, a conserved region (containing an aspartate residue), is hypothesised to be associated with the formation of the phosphorylated intermediate: other functional domains have been identified by inhibitor-binding and site-directed mutagenesis. The postulated mechanism of action of the enzyme involves alternation between phosphorylation and proton binding on the cytoplasmic side of the membrane (E1 state) and dephosphorylation with expulsion of the proton on the outside (E2 state) (see Serrano, 1990; Briskin,

1990). Recent studies with genomic and cDNA clones of *Arabidopsis thaliana*, *Nicotiana plumbaginifolia* and *Lycopersicon esculentum* have indicated that there are multiple forms of the P-type ATPase (see Sussman and Harper, 1989; Ewing *et al.*, 1990). The techniques of molecular biology should, in the future, provide specific antibodies to enable the cytochemical identification of the ATPase and to allow, through the insertion of multiple copies or antisense genes in transformed plants, the effects of changes in the level of activity on the phenotype to be evaluated. Although little is known of how enzyme activity is regulated (Serrano, 1989), it appears that Ca-activated protein kinases (also membrane-bound enzymes) which catalyse phosphorylation may be involved, particularly in regulation by hormones such as auxin. The enzyme in fungi (see Nakamoto and Slayman, 1989) translocates one proton for each molecule of ATP hydrolysed (Sanders, 1988) as is the case for higher plants (see Sussman and Harper, 1989; Briskin, 1990). In the alga *Chara corallina*, the pump may transport one or two protons per ATP hydrolysed (Raven, 1988). In plants the enzyme is central not only to ion transport and the maintenance of turgor, but may also be crucial to growth (the acid growth hypothesis, Rayle and Cleland, 1977). In the acid growth hypothesis it is proposed that transport of protons out across the plasma membrane lowers the pH in the cell wall, and this leads to wall-loosening reactions which allow the cell to expand. Growth is undoubtedly more complex than this, but acidity can, for example, mimic some of the effects of auxins.

3.2.2.2 *V-type ATPases.* Tonoplast or vacuolar ATPases (V-type ATPases or V-ATPases) are also present in the Golgi but, unlike the enzymes of the plasma membrane, are not inhibited by vanadate. From this it may be inferred that the V-type ATPases operate in a quite different manner from the P-type. V-type ATPases also differ from the F_0F_1 type ATPases: they are more sensitive to nitrate and *N*-ethylmaleimide than the F_0F_1 type. The vacuolar ATPases are insensitive to azide and oligomycin, powerful inhibitors of the mitochondrial ATPase and their activity is stimulated by chloride rather than potassium or sodium. The V-type ATPases are large oligomeric proteins with molecular masses of 400–500 kDa, having 8–13 major polypeptide components: three of these appear conserved amongst all the various vacuolar ATPases. These common polypeptides have molecular masses of about 70 kDa, 60 (or 58) kDa and 17 (or 16) kDa. The 60 and 70 kDa subunits constitute part of the hydrolytic centre of the enzyme, although they are not capable of hydrolysing ATP in the absence of the other subunits (Stone *et al.*, 1989).

These subunits are analagous to the α and β subunits of the F_0F_1 ATPases, respectively. The 17 kDa subunit is likely to be a proteolipid making up a channel through which the proton passes across the tonoplast: this subunit binds the proton channel inhibitor, dicyclohexylcarbodiimide (DCCD). There are marked similarities between the V-type ATPases and the F_0F_1 types and they are now considered to have had a common ancestor (Nelson, 1989).

It has been argued that the stoichiometry for the tonoplast ATPase is $2H^+$ transported per ATP hydrolysed (Bennett and Spanswick, 1984), although under most conditions the ATPase is not operating at a thermodynamic equilibrium, but regulated by factors other than ATP supply.

3.2.3 *Pyrophosphatase*

Proton translocation across tonoplast vesicles is not only driven by ATP: there is an equally large component of the activity driven by a pyrophos-

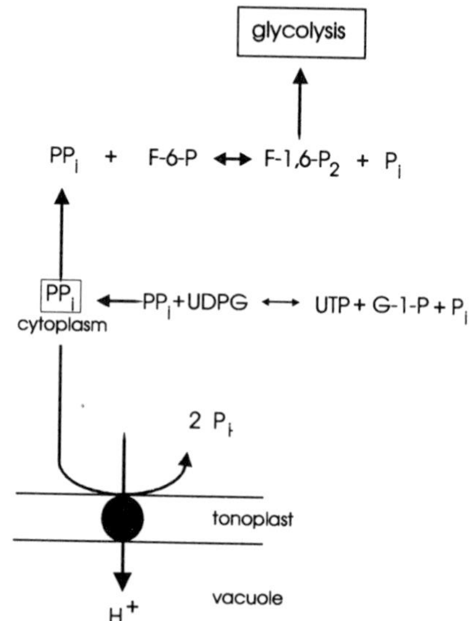

Figure 3.7 The relationships between various origins and fates of pyrophosphate in plant cells. Pyrophosphate (PP_1) can be generated by a variety of reactions (see Taiz, 1986). One of these, which converts glucose 1-phosphate to uridine diphosphate glucose, is illustrated as it is on the route to the synthesis of sucrose. The interconversion of fructose 6-phosphate and fructose, 1,6-diphosphate, catalysed by a phosphofructokinase is reversible, but depicted as working in the direction of glycolysis following Taiz (1986).

phatase (PPase). The discovery of a proton-translocating pyrophosphatase (an enzyme catalysing the hydrolysis of one mole of pyrophosphate to two moles of phosphate) in plants was made in the early 1980s by Walker and Leigh using membrane material isolated from the roots of red beet (see Taiz, 1986). Subsequently, proton translocating pyrophosphatases have been discovered in a wide range of plant vacuolar membranes. PPase activity associated with the tonoplast is generally optimal at pH values of 7.5–8 and is greatly stimulated by KCl and RbCl, but unaffected by vanadate or nitrate.

There are other well known pyrophosphatases, particularly soluble cytoplasmic enzymes, present in plant cells that catalyse the interconversion of pyrophosphate and inorganic phosphate. These soluble enzymes have been commonly held to give direction to a number of anabolic reactions that result in the synthesis of pyrophosphate, by hydrolysing the pyrophosphate. The discovery of the tonoplast H^+-translocating pyrophosphatase and a pyrophosphate-phosphofructokinase has lead to the suggestion that these two enzymes utilise the bond energy in the pyrophosphate generated in anabolic reactions and provide a direct coupling of biosynthetic reactions with transport across the tonoplast (Taiz, 1986; see Figure 3.7).

3.3 Redox systems

Although the proton-pumping enymes are thought to be the prime agents in electrogenic transport in plant cells, there are also other mechanisms that might move charge across membranes. The presence of redox systems in the plasma membrane of mammalian, bacterial and yeast cells is now firmly established and similar systems are present in the plasma membrane of plant cells (Crane and Barr, 1989). There is, however, little evidence concerning tonoplast membranes.

Purified plasma membrane from plants has been shown to be able to oxidise reduced nicotinamide adenine dinucleotide (NADH) and nicotinamide adenine dinucleotide phosphate (NADPH)—the two co-enzymes are conveniently abbreviated as NAD(P)H—as well as being able to reduce cytochrome c and ferricyanide (to ferrocyanide), when presented with NAD(P)H. These reactions involve a number of redox components, present in the plasma membrane: *b*-type cytochromes, peroxidase(s) and flavins (see Møller and Crane, 1990). The exact role of the redox systems in solute uptake by plant cells is, however, still far from certain. While it is clearly possible that such a system could be electrogenic, by moving charge or protons across the membrane, it seems unlikely that the redox system does extrude protons. If anything, it is more likely that redox reactions act by depolarising the membrane potential (by exporting negative charge) and thereby stimulate the activity of the P-type ATPase by reducing the free energy gradient against which it is acting. While the possible functions of the redox systems remain a matter of speculation their one established role is to reduce external iron. It is in the ferrous state (Fe^{2+}) that iron is taken up by the roots of most plants through ion channels for divalent cations. Ferric iron (Fe^{3+}) rapidly forms hydroxides and other insoluble salts.

Surprisingly, in monocotyledonous species iron is generally taken up in the ferric state by complexing to an organic molecule that has been secreted by the roots (see Marschner, 1986). These molecules are samples of a group of compounds produced by plants and bacteria, that are able to chelate ferric iron and known as *siderophores* (from the Greek word for iron bearers); those compounds secreted by plants are known as *phytosiderophores*. They are hydroxy- acid and amino-substituted imino carboxylic acids (Neilands and Leong, 1986): mugineic acid is secreted by wheat and avenic acids by oats. In some species additional capacity to reduce external iron is induced under iron deficiency: it is known as the 'turbo system'. This turbo system may be capable of making new sources of iron available through its ability to reduce iron in phytosiderophores (see Crane and Barr, 1989).

3.4 The transport of ions other than protons

The picture that is emerging at the cellular level of ion transport in plants is one where the movement of protons is pivotal to the movement of other solutes: at both plasma membrane and tonoplast proton transport is electrogenic. The total membrane potential (including diffusive components) provides a free energy gradient for the entry of cations into the cytoplasm. However, although there is an enormous literature on the phenomenology of ion transport in plants, we still know very little of just how ions enter the cells of higher plants and are stored in their vacuoles. Much of what we do know about has come very recently, from studies using the patch-clamp technique (section 2.5.1).

3.4.1 *Potassium*

For most plants, the only cation that is required internally at relatively high concentrations is potassium; to activate enzymes in the cytoplasm and to generate the osmotic potential necessary for cell expansion. This potassium is likely to enter the cells passively, crossing the plasma membrane via an ion channel or channels. The membrane potential can usually account for much, even all, of the potassium accumulation, i.e. the potassium is in approximate equilibrium across the plasma membrane as determined by the Nernst equation. The entry of potassium may be a consequence of the membrane potential generated by electrogenic processes occurring at the plasma membrane.

The presence of potassium channels in the membranes of plant cells was first suggested by analysis of the action potentials that occur in charophytes (algae such as *Chara* and *Nitella* in the Charophyceae, a class within the Chlorophyta). Subsequently, channels that open on depolarisation, that is as the membrane potential rises from a negative value to become more positive than -40 mV, have been shown to be widespread in the plasma membrane of plant cells (see Hedrich and Schroeder, 1989; Bentrup, 1990; Tester, 1990). These channels favour the outward movement of potassium from the cell (and are consequently known as *outward rectifiers* or as $I_{K^+,out}$ channels). They are more permeable to potassium than rubidium or sodium and those from pulvini (a pulvinus is a group or cushion of parenchyma cells, generally at the base of a petiole, which is responsible for movements of the leaf; *Mimosa pudica* is a well known example of a species showing striking leaf movements) are antagonised by the classic potassium channel blocker, tetraethylammonium (see Hedrich and Schroeder, 1989). There is also another type of potassium channel present in the plasma membrane and this appears to facilitate the inward movement of potassium (hence *inward rectifiers*, $I_{K^+,in}$). These channels are activated by membrane potentials more negative than -100 mV and may be the major path for potassium uptake by cells. Such potentials would presumably derive, at least in part, from the action of the plasma membrane ATPase.

An interesting question arises with respect to how potassium stimulates the activity of the plasma membrane ATPase. Is this through the release of a constraint on proton pumping, the constraint (a high ΔH^+) being dissipated via an ion channel (that is, effectively, an H^+/K^+ antiport), or due to a direct effect of potassium on the enzyme activity itself (Sze, 1985)?

Potassium does stimulate the turnover of phosphorylated protein (see Briskin, 1990) and could at this time be translocated through the membrane, as with the animal cell Na/K ATPase. The plant plasmalemmal ATPase may even have a potassium channel as part of the native enzyme, which is involved in potassium uptake by the cell (Serrano, 1989). The current consensus, however, is that the P-type ATPase operates as a proton uniport and potassium enters through independent K^+/H^+ symport (see Bentrup, 1990) or K^+ uniport (see Briskin, 1990).

As far as the tonoplast is concerned, there is also evidence from many plant species of ion channels that conduct potassium (Tester, 1990). The vacuolar membrane has inwardly (into the vacuole) directed currents of two types. There are inwardly rectifying channels which are slowly (hence slow-vacuolar or SV-type currents) activated by calcium ($> 0.3\,\mu M$) with permeability to both anions and cations: they are not selective between cations such as sodium and potassium, for example. There are also fast-vacuolar (FV) types, activated by low ($< 0.3\,\mu M$) calcium, which it has been suggested may provide a means by which anions accumulate in the vacuoles and potassium equilibrates between cytoplasm and vacuole (Hedrich and Schroeder, 1989).

3.4.2 *Nitrate and chloride*

Little is known of the means by which the other ions present in, and necessary for the growth of, higher plants enter their cells. Only nitrogen, which is present in the soil as the ammonium cation (NH_4^+) or nitrate anion (NO_3^-), is required in amounts similar to potassium. The ammonium ion can cross the membrane freely as the uncharged NH_3 molecule (cf. section 2.6.1). Nitrate is absorbed in considerable quantities by most plants but as the cells are electronegative inside with respect to their surroundings, entry of anions into the cell will be against their free energy gradient in practically all circumstances. This being so, NO_3^- may enter in some form of symport with protons.

The only other anion for which there is a lot of information on its transport is chloride. In the alga *Chara*, there is a $2H^+/1Cl^-$ symporter, one of only two proton cotransport systems definitively identified: the other is a $1H^+/1$ glucose symporter in *Chlorella* (see Raven, 1988). Chloride fluxes are important in action potentials, which some plants, particularly Characean algae, display. A stimulus of some kind, perhaps mechanical, causes a transient depolarisation of the membrane potential. The potential falls to about $-30\,mV$ and then returns to the resting potential of about

−170 mV over a period of about 3 s: the physiological function of the action potential is unknown. During this depolarisation, negative charge leaves the cell and is carried by an efflux of chloride ions (see Lüttge and Higinbotham, 1979). A chloride channel which appears to be activated by a rise in cytoplasmic calcium (see Tester, 1990) is present in the Characean plasma membrane and in the tonoplast. Another chloride channel which is present in the plasma membrane of *Chara* opens on hyperpolarisation. A further channel activated by stretching of the membrane has been suggested and might be part of a mechanism that senses turgor pressure and hence is involved in osmotic adjustment.

In the alga *Acetabularia* (Ulvophyceae), chloride influx is effected by primary active chloride transport powered by ATP, utilising 2ATP per Cl^- transported and sensitive to vanadate (see also section 3.1). A primary active chloride influx pump that is electrogenic also appears to be present in other orders of the algae (see Raven, 1988).

3.4.3 *Calcium*

Calcium is, after potassium, the most abundant cation in plants: overall calcium is commonly present at concentrations of 1 to 10 mM, although its concentration within the cytoplasma is very low (30 to 400 μmol m^{-3}; Table 1.4). In the cytoplasm calcium must be at a low concentration because of its toxicity, which is due to its ability to precipitate phosphates and hence interfere with the energy metabolism of the cell. The low cytoplasmic concentration means there is generally a strong inward driving force for calcium and efflux from the cytoplasm is always likely to be against a free energy gradient. Mechanisms must clearly exist to regulate such low concentrations especially since calcium itself seems to play an important role in the regulation of many processes in both plant and animal cells (see Hepler and Wayne, 1985 and section 4.1).

There is evidence for an ATP-dependent primary calcium transport system (a Ca-ATPase associated with Ca transport) in vesicles prepared both from the endoplasmic reticulum and the plasma membrane; both ATPase activity and transport are stimulated by calmodulin, a calcium binding protein found both in animal and plant cells (although not all preparations show stimulation, see Evans *et al.*, 1991). The Ca-ATPase is a P-type ATPase with a molecular mass of about 130 to 140 kDa, although there is evidence that the functional enzyme is a dimer. There are conspicuous similarities between the Ca-ATPase from maize and that from red blood cells (see Evans *et al.*, 1991).

Channels that conduct calcium have also been detected in the plasma membrane of charophytes and higher plants (see Johannes *et al.*, 1991). These channels differ in their sensitivity to the antagonists, verpamil and 1,4-dihydropyridine. A protein that is likely to be a calcium channel has been partially purified from maize coleoptiles (see Tester, 1990). The calcium channel is an inward rectifier that is permeable to monovalent cations when divalent cations are present in very low (cytoplasmic) concentrations, a characteristic of many calcium ion channels in animal cell membranes. It is likely that gating of this channel is controlled by phosphorylation.

In the tonoplast there is evidence for an inositol-1,4,5-triphosphate stimulated calcium channel, analogous to that present in the endomembranes of animal cells, (Tester, 1990). The tonoplast has also been shown to contain a Ca^{2+}/H^+ antiport: the V-type ATPase and the PPase generate the proton gradient which provides the energy to accumulate calcium in the vacuole against its concentration (and chemical potential) gradient—vacuolar concentrations of calcium are commonly up to 10 mM. A range of calcium channels, which include a variety of voltage gated channels, showing a 1000-fold selectivity for calcium over sodium or potassium are to be found in the membranes of animal cells (Tsien and Tsien, 1990).

3.4.4 *Magnesium and minor nutrients*

After calcium, magnesium is probably the next most abundant cation in cells, where it plays a central role in the activation of many enzymes as well as being part of the substrate for the proton translocating ATPases (the substrate is Mg-ATP). The concentration of magnesium in the cytoplasm is about 3–4 orders of magnitude higher than that of calcium. However, in spite of the importance of magnesium, little is known of its mode of entry.

Other nutrients such as boron, iron, manganese, zinc, copper and molybdenum are required at low concentrations, mostly as part of a metalloprotein or pigment, or performing specific functions as cofactors in enzymic reactions (see Clarkson and Hanson, 1980; Marschner, 1986). Again little is known of their mode of entry into the cell. Presumably, however, these cations enter either through specific channels or via carriers.

Although there is a vast amount of literature describing many aspects of the physiology of ion transport in plants, dating back to the turn of the century, the interpretation of this data in a modern context is still in its infancy. Two examples of specialised transport systems are now presented to illustrate the processes occurring; others follow in later chapters.

3.5 Guard cells

Stomata open and close due to changes in the turgor pressure within the guard cells. During stomatal opening, an increase in the concentration of solutes within the vacuoles lowers the free energy of water in the guard cells and causes water to enter from neighbouring epidermal cells. This increases turgor pressure and generates forces in the walls that bring about stomatal opening (see Willmer, 1983). Closure of the stomata is, essentially, the reverse of opening; the vacuoles lose solutes and water and the cells turgor. The vacuoles of guard cells are atypical amongst those of higher plants in that they undergo diurnal changes in their salt content and their volume: the vacuoles in most plant cells remain with a relatively constant solute concentration throughout their lives.

While the measurement of stomatal aperture using a microscope is a relatively simple procedure that can be correlated with direct measurements of conductance to gases, it is much more difficult to determine the solute fluxes that bring about changes in aperture. Much of the work that has been undertaken on solute movements has concentrated on two species, *Vicia faba* (broad bean) and *Commelina communis* (day flower), because of the ease with which their guard cells can be observed. Particularly important has been the use of epidermal strips, where all but the guard cell protoplasts can be killed. Cells isolated from the intact tissue do differ from those *in situ*, however, in that they require smaller changes in solute concentration to effect the same change in volume because there is no back-pressure from neighbouring epidermal cells.

There has been much hypothesising about the solutes that change in concentration to effect changes in volume. Both sugars and salts have, in the past, been ascribed a central role (see Willmer, 1983). Currently, potassium is seen as the major cation involved, with chloride and the organic anion, malate, being responsible for balancing charge. The fact that changes in potassium concentration occur within the guard cells during stomatal opening and closing is well documented (MacRobbie, 1988). In *C. communis*, the change in concentration is from about 500–900 mM in the open state to around 80–110 mM in the closed state (Figure 3.8).

The net fluxes of potassium (with chloride and malate) associated with changes in stomatal aperture can be estimated from the time taken to open and close, the volume change needed to achieve this, and the measured surface areas of the guard cells. As the guard cells are essentially isolated from cells other than their neighbouring epidermal cells, these fluxes must occur across the plasma membrane. Since it takes a net movement of about

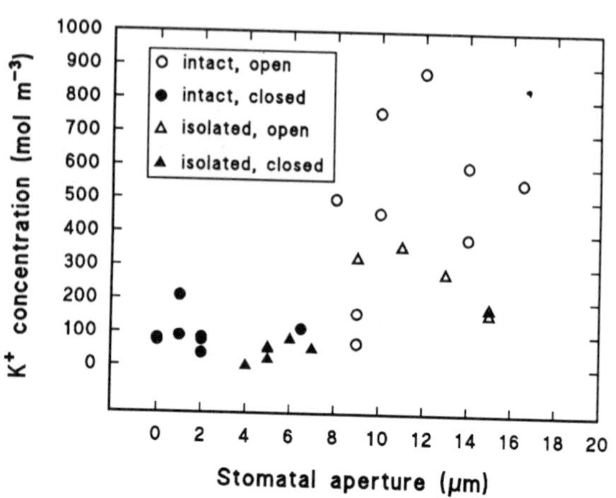

Figure 3.8 The relationship between the potassium concentration in the guard cells of a variety of species (*Vicia faba, Commelina communis, Nicotiana tabacum* and *Allium cepa*) and the stomatal aperture. Data are for guard cells intact in tissues and those isolated in epidermal strips. Drawn from data in Table 12.1 in MacRobbie (1988).

1.2 mmol of potassium per m^2 of guard cell surface area in both *V. faba* and *C. communis* for opening and closing and 1–2 h for the stomata to open but only 0.3 h to close, the net closing flux is much higher than the opening flux: for opening (over 90 min) the flux would be 220 nmol m^{-2} s^{-1}, whereas for closing the flux is 1000 nmol m^{-2} s^{-1}. These are net fluxes, the sum of the influxes and effluxes occurring across the plasma membrane and tonoplast, and analysis of the movement of tracers (cf. section 1.2.2) into and out of the cells has been used to investigate the effects of the opening and closing signals on the individual fluxes across plasma membrane and tonoplast (see MacRobbie, 1988).

Measurements on the effects of fusicoccin, which stimulates stomatal opening, show that opening in *C. communis* involves an increase of influx and a reduction of efflux. These fluxes are strongly influenced by calcium and there is a view that stomatal aperture might be regulated through the cytosolic calcium concentration (see Mansfield *et al.*, 1990). The potassium flux has been postulated to occur through an ion channel ($I_{K,in}$; Hedrich and Schroeder, 1989). Patch-clamp studies of *V. faba* guard cell protoplasts have demonstrated potassium-channel activity sufficient to allow the flux required for opening, i.e. 200 nmol m^{-2} s^{-1} with a membrane potential 30 mV away from the equilibrium potassium potential and one channel per 15 μm^2, open one third of the time. There are two lines of evidence to connect the potassium fluxes to the action of the ATPase in the plasma membrane. First, a short (1 to 100 s) pulse of bule light, which stimulates transient stomatal opening, has been shown to bring about H^+ efflux from *V. faba* protoplasts. Second, fusicoccin, as well as stimulating stomatal opening, stimulates the plasma membrane ATPase. In *V. faba*, proton fluxes of 30–100 nmol m^{-2} of guard cell area per second are of a similar order to the fluxes of potassium. Blue light is clearly an activator of the proton pump, although the connection between a blue light receptor and the P-type ATPase is not clear.

Closing, induced both by darkness and the hormone abscisic acid (ABA), involves the activation of the excretion of salt through a transient stimulation of efflux: influx is not inhibited during the closing response (MacRobbie, 1988). This efflux may take place through the $I_{K,out}$ channel mentioned earlier (section 3.4.1; Hedrich and Schroeder, 1989).

Apart from the interesting situations of opening and closing, there are also two other states in which the ion fluxes must be regulated: the turgid, open state and the less turgid, closed state. In these steady states, fluxes across both tonoplast and plasma membrane are lower when the stomata are closed than when they are open (MacRobbie, 1988).

Although the discussion so far has concentrated on the movement of potassium ions, there must also be transport of a counter ion—chloride or malate. Any chloride influx across the plasma membrane is likely to be active, as the electrochemical potential of chloride in the electronegative cytoplasm will be higher than that in the cell walls. The mechanism of this transport process in unknown in guard cells, but should it be similar to that in the giant alga *Chara*, where there is a $2H^+/Cl^-$ cotransport system (see section 3.4.2), with a proton pump moving $1H^+$ per ATP hydrolysed, then the influx of one Cl^- would require the hydrolysis of two ATP. MacRobbie (1988) has calculated that this would require virtually all the ATP synthetic capacity of the guard cell protoplasts.

3.6 CAM plants

In all plant species the net fixation of carbon relies on the activity of the enzyme ribulose bisphosphate carboxylase (this is given the acronym Rubisco because the enzyme acts as both carboxylase and oxygenase) and the Calvin cycle (see also section 4.2). In some species, however, carbon is also fixed by phosphoenolpyruvate carboxylase (PEPC), producing oxaloacetate, which is subsequently reduced to malate, catalysed by the enzyme malate dehydrogenase (Figure 4.7). The particular significance of the production of malate is that it can subsequently be decarboxylated (by another malate dehydrogenase) to release carbon dioxide where needed, at the site of Rubisco activity. In C_4 plants, this process occurs in the light, during photosynthesis. The two carboxylases are spatially separated and there is a flux of malate (the currency of carbon transfer) between the sites of PEPC and of Rubisco (see also section 4.2.). The process appears to have evolved as a means of increasing the amount of carbon fixed per unit of water lost in transpiration in order to increase the water use efficiency of carbon assimilation in plants of hot, dry environments. In yet other species, typified by members of the Crassulaceae, the process has become even more extreme. In these plants, the stomata open during the night, when the evaporation rate for water is lower than during the day. By this means water can be conserved but at the cost of preventing the uptake of carbon dioxide from the atmosphere during the photoperiod. In such plants carbon fixation is achieved by using PEPC and Rubisco in a process where the activities of the two enzymes are separated temporally: PEPC is primarily active during the night and Rubisco during the day. In such plants, the carbon dioxide is fixed into malate during the dark period and stored in the vacuoles. During the day, carbon dioxide is released from the malate (by the decarboxylating malate dehydrogenase, also known as malic enzyme) to provide the carbon dioxide for the Calvin cycle which can operate in spite of the fact that the stomata are closed. Plants operating this process are known CAM plants (Crassulacean acid metabolism). In both C_4 plants and CAM plants there is a substantial traffic in malate. In CAM plants this is dominated by massive acidification during the night and de-acidification during the day: the rhythm of titratable acidity typifies CAM. As far as transport processes are concerned, malate must be transported across the tonoplast in both directions.

There is, then, a massive diurnal change of about 200 mM in the concentration of malate in the vacuoles of CAM plants. This is reminiscent of the change in solute concentration in the vacuoles of guard cells. However, an important difference between guard cells and CAM plants is that the movement of anion (malate) is accompanied by protons in the CAM plants, not by any other cation. Hence in the CAM plants the vacuolar acidity increases during the night from about pH 6 to pH 3.3. The stoichiometry is that two protons are accumulated with each malate. Malate is, of course, a dibasic acid therefore the free acid ($MalH_2$) can dissociate to release one proton and $MalH^-$ or two protons and Mal^{2-}. From estimates of the malate concentration in the vacuoles and their pH, and estimates of the malate concentration in the cytoplasm as well as its pH, it can be calculated that only Mal^{2-} is distributed passively across the tonoplast according to the Nernst relationship (Lüttge and Smith, 1988). Protons or $MalH^-$ would have to be actively transported. There is no evidence for the active transport of $MalH^-$, whereas there is good evidence for the active transport of protons.

Vacuoles isolated from the leaf cells of a typical CAM plant, *Kalanchoë daigremontiana*, utilise ATP in the transport of protons. Proton pumping is typical of transport catalysed by V-type ATPases, because it is insensitive to vanadate and inhibited (about 40%) by nitrate. ATPase activity also reflects CAM activity. In some species, such as *Mesembryanthemum crystallinum*, CAM can be induced by drought or salinity. When induced by salinity, sodium chloride increases in the vacuoles before the induction of CAM. Only on the induction of CAM does ATPase activity rise. However, in spite of this clear indication of the role of the ATPase in CAM, very little is known of the regulation of enzyme activity during the diurnal cycle or of the means by which malic acid crosses the tonoplast.

3.7 Bacterial and yeast systems

The study of solute transport in plants has generally lagged behind similar studies in other organisms and so knowledge of these other transport systems may be useful in suggesting directions for future research on plants. Any review of solute transport in animals is, however, clearly beyond the scope of this book as would be thorough reviews of transport phenomena in bacteria and fungi. For the latter two groups, however, some comment is pertinent as there appear to be many similarities between the transport systems in plants, bacteria and fungi.

The real advantage to be gained in the study of transport in bacteria and fungi is not in overcoming the technical difficulties of working with plants, but in the ease with which genetic studies can be undertaken. For most plants the generation time is measured in months or years so genetic experiments are particularly long-term affairs. For bacteria and fungi (such as yeasts) the generation time is measured in hours. The rapidity with which generations can be screened and the relative ease of genetic manipulation of these predominantly haploid organisms conveys a significant advantage to their use in the evaluation of the role of particular proteins in the organism. For example, it should be possible to evaluate the importance of a P-type ATPase by modifying the genome to produce either an incompetent protein or to produce multiple copies of the enzyme and determine the consequences for the organism as a whole (cf. section 3.2.2.1).

3.7.1 *Bacteria*

There is a vast amount literature on ion transport in bacteria, but recent studies have confirmed the central importance of primary active proton transport systems and proton cotransport in non-marine organisms. Interestingly, marine and halophilic bacteria appear to use Na^+ in place of H^+ both in primary active transport and in coupled transport. Primary active sodium transport may be brought about by Na-ATPases, Na-translocating decarboxylases and respiration-dependent Na^+ pumps (see Rosen, 1986). Subsequently, sodium exchange can occur through a Na^+/H^+ antiport via a monovalent cation/proton antiporter or a specific Na^+/H^+ antiporter. Potassium translocating ATPases are also known with distinct homologies to the proton translocating P-type ATPases.

Within the Gram-negative bacteria (which include that much studied organism, *Escherichia coli*) there is a particularly interesting group of ion channels to be found in the envelope that surrounds the organism. This envelope is three-layered, consisting of an outer membrane, a peptidoglycan layer and an inner membrane. The outer membrane contains a class of channels known as *porins*. They can be extracted from the membrane, have been purified and their properties studied in artifical membranes. The porins have a molecular mass of 30–50 kDa, exist as trimers and form a rather large channel in the membrane, with a diameter of about 1.2 nm. They allow a range of quite large solutes to permeate the outer membrane, but have some selectivity towards anions and show characteristics of voltage gating (Benz, 1985). As far as plants are concerned, the major interest in porins lies in the presence of physiologically, but not chemically, similar channels in the outer membrane of chloroplasts and mitochondria (see sections 4.1 and 4.2).

3.7.2 *Fungi*

The fungi are an enormously diverse group, but two Ascomycetes, *Saccharomyces cerevesiae* and *Neurospora crassa* have been used extensively in studies of solute transport. Evidence both for proton pumping and ATPase activity at the plasma membrane of *Neurospora crassa* preceded that in higher plants (see Sanders, 1988). The ATPase has been purified: electrophoresis in the presence of the detergent sodium dodecylsulphate

(SDS) shows a monomeric molecular mass of 100 kDa. Its specific activity, which is sensitive to vanadate, is 100 μmol mg^{-1} protein per minute, with a turnover number of 150 per second assuming that the monomer is the functional unit (although claims have been made that a dimer is necessary). The enzyme, which crosses the membrane eight times (cf. section 3.2.2.1), pumps a single proton per ATP hydrolysed: it appears to be very similar to the P-type ATPases found in the plasma membrane of higher plants. The vacuolar ATPase is also similar to that isolated from higher plants, with a molecular mass of 530 kDa.

Fungi are able to accumulate a wide range of solutes from their external medium, from sugars in various forms through amino acids, purines and pyrimidines, to inorganic ions. They have low-affinity systems that are mainly constitutive and also inducible systems (see Sanders, 1988). The primary active transport system for uptake of these solutes is proton efflux. The uptake of the solute generally involves proton cotransport.

CHAPTER FOUR
INTRA- AND INTERCELLULAR TRANSPORT

Although the transport of ions across plasma membrane and tonoplast is central to many aspects of plant life, transport is not confined to these membranes. Ions, together with other solutes, move into and out of a variety of subcellular compartments. In this chapter we will consider the transport of solutes within the cytoplasm and between neighbouring cells, i.e. local movement. This involves not only fluxes of ions between subcellular compartments and cells, but also traffic in metabolites, including carbon fixed during photosynthesis. Solute transport over longer distances is considered in the later chapters.

As has already been emphasised, the largest single compartment in a mature plant cell is generally the vacuole, bounded by the tonoplast. Other important compartments are made up by the organelles, the chloroplasts in photosynthetically active cells and the microbodies and mitochondria in all cells. Although they may be small by comparison with the vacuoles, these compartments contain most of the cell's complement of particular metabolites. Transport across the membranes of mitochondria and chloroplasts is additionally interesting because of links to bacterial transport systems, reflecting the possible evolutionary origins of these organelles (Margulis, 1981). It is in the membranes of mitochondria and chloroplasts that carrier proteins assume a singular importance.

4.1 Mitochondria

Mitochondria are ubiquitous in the cells of higher plants. These organelles, which are about 3–5 μm long and 0.5–1.0 μm in diameter, vary in number in different cells and different plants; from about 20 to 10 000 per cell (Hall *et al.*, 1982). Mitochondria are relatively easy to isolate using differential centrifugation and can be obtained in a fairly pure state from all sorts of plant tissues (see Douce and Day, 1985). Each mitochondrion has an envelope comprised of two separate membranes. The *outer membrane*

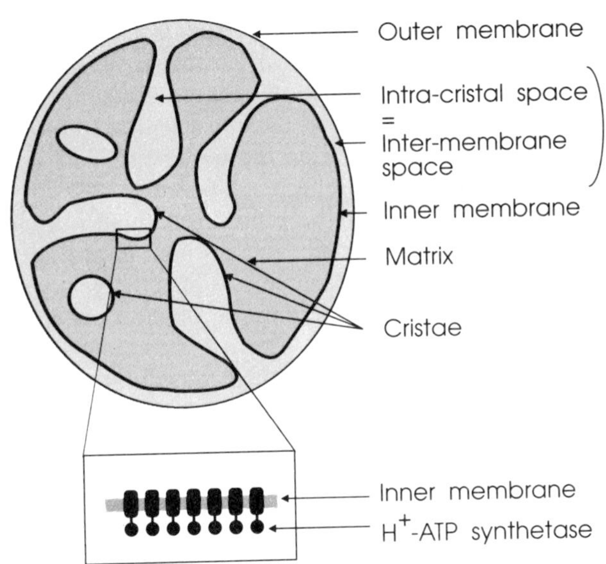

Figure 4.1 A diagrammatic representation of the structure of a plant mitochondrion showing the cristae, which are convolutions of the inner membrane and the sites of the ATP synthetase.

surrounds the organelle, while the *inner membrane* is folded into the lumen to form so-called *cristae* (Figure 4.1). The inner membrane and its cristae separate the inner compartment of the mitochondrion, the *matrix*, from the intracristal space: this space, between the invaginations of the inner membrane, is contiguous with that between inner and outer membranes and the whole is known as the *intermembrane space* (Figure 4.1).

The mitochondria are the site of the enzymes of respiration, and the synthesis of ATP, which is subsequently exported. Plant mitochondria may also export a considerable proportion of the carboxylic acids produced during the operation of the tricarboxylic acid (TCA) cycle; these carbon skeletons are used for anabolic reactions taking place elsewhere in the cell. Mitochondria are also the site of synthesis of serine from glycine. Consequently, mitochondrial transport systems must mediate the import of inorganic phosphate (P_i) and adenosine diphosphate (ADP), the export of ATP and the import and export of carboxylic acids and some amino acids.

The transport of substances into and out of mitochondria is a little easier to investigate than is transport into and out of membrane vesicles. The mitochondria are considerably larger than the vesicles and transport data is easier to interpret than with vesicles formed from a mixture of cell membranes: the mitochondrion is a clearly recognisable organelle with a defined inside and outside. It is also possible to measure, optically, the swelling and contraction that follows the movement of solutes across the mitochondrial envelope by measuring the transmission or scattering of light. The uptake and loss of radionuclide-labelled probes can be readily determined as mitochondria can be centrifuged rapidly from reaction media through a layer of silicone oil, which is used to prevent mixing of reaction and collection media. Measurements of activity (oxygen uptake) can also give information on the permeation of substrates (see Hanson, 1985).

Much has been learned of the movement of solutes across the mitochondrial membranes from measurements of the optical density of mitochondrial suspensions: changes in absorbance at 520–550 nm are used to monitor changes in volume following solute and water movement (Hanson, 1985). The outer membrane is permeable, even to molecules with molecular masses of 5000–10 000. This permeability is due to the presence of channel-forming proteins similar to the porins of the outer membrane of the envelope of Gram-negative bacteria (see section 3.7.1). These channels, which are also found in mammalian and yeast mitochondria, are formed by a polypeptide of 30–35 kDa. The channels, which are voltage gated, are more permeable to anions than cations, hence they are sometimes known as voltage dependent anion-selective channels (VDACs). The operational channel is an intrinsic protein composed of one or two polypeptides (Manella and Tedeschi, 1987). VDACs have a clear functional similarity to the bacterial porins (which might suggest a similarity in the sequence of amino acids, given the possible evolutionary origins of mitochondria; Margulis, 1981), but nothing in the sequence of amino acids in VDACs from yeast suggests they are related to those in bacteria (Forte *et al.*, 1987).

While the outer membrane of the mitochondrial envelope is permeable to molecules such as sucrose, changing the external sucrose concentration brings about changes in the volume of mitochondria. This is evidence that sucrose does not readily permeate the *inner* membrane. Changes in volume are effected across this inner membrane and the mitochondria behave as near perfect osmometers. Changes in volume can also be caused by changing the nature of the external solution. For example, if plant mitochondria that have been previously suspended in buffered sucrose are transferred to buffered potassium chloride (about 100–200 mol m^{-3}), they swell spontaneously (Figure 4.2). Since under the circumstances, the potassium concentration within the matrix is about 120–140 mol m^{-3},

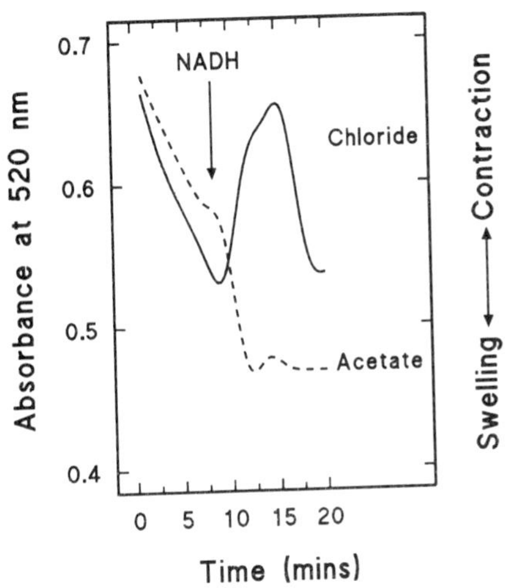

Figure 4.2 Swelling and contraction in plant mitochondria. The diagram shows the change in optical density measured at 520 nm, which reflects the volume of the mitochondrion, with time. On transfer from sucrose to potassium chloride (at time = 0), the mitochondria swell. Contraction is induced by the addition of a small amount of respirable substrate (NADH). Once this has all been respired, the mitochondria re-swell. In potassium acetate, as opposed to potassium chloride, the addition of NADH promotes further swelling rather than contraction (see text for explanation).

there is little gradient in potassium across the inner membrane when the mitochondria are transferred to KCl (say 150 mol m^{-3}). It is thought that chloride enters down its concentration gradient since within plant mitochondria suspended in sucrose, chloride is only about 10 mol m^{-3}. Chloride entry is electrogenic, generating a membrane potential across the inner membrane, negative within the matrix, which provides the driving force for the entry of K^+. This K^+ may then exchange with H^+ via a K^+/H^+ antiport. The increase in KCl concentration in the matrix lowers its water potential (section 2.8.2) and water enters, making the mitochondria swell. The potential differences (water and electrical) and concentration gradients between the matrix and the cytoplasm are generated across the inner, and not the outer, membrane of the mitochondrion.

The inner membrane is crucial to the most well known aspect of mitochondria, the synthesis of ATP. This has occupied the centre stage in mitochondrial biochemistry for many years. Early debate about the mechanism of ATP synthesis has given way to acceptance of the chemiosmotic hypothesis first elaborated by Mitchell (1966). Membrane transport of protons and an ATPase are essential features of this theory.

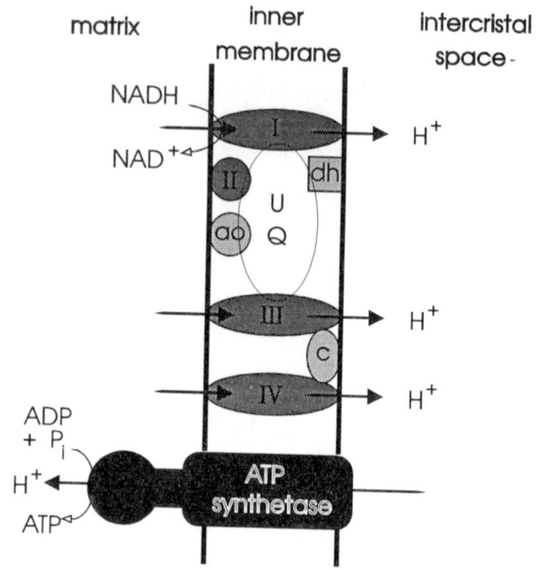

Figure 4.3 The respiratory chain and ATP synthetase in the inner membrane of a plant mitochondrion. The respiratory complexes (I, II, III and IV) are linked by ubiquinone (UQ) and cytochrome c (c). NADH from the TCA cycle enters at complex I and succinate at complex II. Oxidation of reduced substrates results in the net transfer of protons from the matrix to the intracristal space. ATP synthesis is driven by the pmf generated. The alternative oxidase (ao) and the NADH dehydrogenase (dh) accessible from the outside of the mitochondrion are also illustrated.

However, in contrast to the other ATPases that have been described so far, this enzyme is an ATP synthetase; ATP is normally synthesised not hydrolysed.

The energy of a proton gradient, generated across the inner mitochondrial membrane by the action of the respiratory chain (Figure 4.3), is used in the production of ATP from ADP and P_i. The arrangement of the components of the respiratory chain within the inner membrane brings about the transfer of protons from the matrix into the intermembrane space. This proton transport is electrogenic, producing what has become known as the proton motive force (pmf). It can be quantified by calculating the electrochemical potential (see section 2.8.1) difference for protons across the membrane, viz.:

$$\Delta\mu_{H^+} = \mu_H^i - \mu_H^o = (RT \ln a_H^i + zFE^i) - (RT \ln a_H^o + zFE^o)$$

$$= RT \ln \frac{a_H^i}{a_H^o} + zF(E^i - E^o)$$

where i and o represent the two sides of a membrane, a_H is the activity of protons, z is the valency (1, for H^+) and E the electrical potential: the other symbols have their previously defined meanings.

$\Delta\mu_{H^+}$ can be represented as a difference of pH and electrical potential across the membrane, by substituting the term in activity for one in pH. This is easily done, since

$$pH = -\log a_H = -\ln a_H/2.303,$$

whence

$$\Delta\mu_{H^+} = -2.303\,RT(pH^i - pH^o) + zF(E^i - E^o)$$

At 20°C, the expression reduces to:

$$\Delta\mu_{H^+} = 5.6(pH^o - pH^i) - 96.5(E^o - E^i) \qquad kJ\,mol^{-1},$$

where E is in volts. The equation relates the pmf to ΔpH and ΔE.

Protons pumped into the intermembrane space by the respiratory chain re-enter the matrix down their electrochemical potential gradient through a channel in the inner membrane that is part of the ATP synthetase. The enzyme, which can be seen as a stalked protein protruding into the matrix from the inner mitochodrial membrane, is responsible for coupling the transport of protons to the synthesis of ATP. It is an F_0F_1 type ATPase, consisting of up to 12 subunits and a total molecular mass of about 500 kDa (Table 3.1). The proton channel is made up of the F_0 protein (which consists of three subunits. *a*, *b* and *c*) and this may be blocked by the inhibitor, DCCD which binds to subunit *c* (see also section 3.2.2.2). The stoichiometry is such that either two or three protons are translocated for each molecule of ATP synthesised.

In spite of the importance of proton transport across the inner mitochondrial membrane for the synthesis of ATP, plant mitochondria have, in practice, a very small difference of pH across this membrane. There is, however, a considerable electrical potential difference of 220 mV which is negative on the matrix side (see Douce and Neuburger, 1989). The free energy gradient of protons, the pmf, is thus comprised primarily of an electrical component (ΔE); the chemical component (ΔpH) of the electrochemical potential difference of protons across the membrane is small. This can be explained by the presence of a K^+/H^+ antiporter which is electrically neutral. This antiporter exchanges protons for potassium ions, thus dissipating the pH gradient, while maintaining the electrical gradient. If protons were not rapidly returned to the matrix, the supply in this small compartment would be depleted rapidly.

Although the synthesis of ATP is driven by the electrochemical gradient of protons across the inner membrane, the pmf can also be used to drive the transport of other substances, e.g. in secondary active transport in a way similar to that already described for the plasma membrane and tonoplast. On addition of a source of energy, a respirable substrate such as NADH for example, the mitochondria contract (Figure 4.2). This is the consequence of the extrusion of protons from the matrix during respiration, followed by the efflux of Cl^- driven by the change in membrane potential and H^+/K^+ antiport. There is a net efflux of KCl, an *increase* in the internal water potential, establishing a water potential gradient with the cytoplasm, from which follows the outflow of water and mitochondrial contraction. In contrast, with acetate as the counter ion, there is energy dependent swelling (Figure 4.2), as acetic acid (acetate$^-/H^+$ symport) enters and the cation follows due to the very negative potential in the matrix. However, such swelling does not occur with calcium (Hanson, 1985).

Calcium uptake into mitochondria in mammalian cells is an energy dependent process and may play a role in the regulation of cytoplasmic calcium ion concentrations. Plant mitochondria also take up calcium in an energy dependent process, although at considerably lower rates than animal mitochondria. There may be substantial differences between different plant species and perhaps even tissues. Little is known of the specific mechanism involved, although calcium uptake by plant mitochondria is dependent on phosphate uptake. Even less is known of the uptake of other inorganic ions important to the mitochondrial metabolism, such as magnesium or iron: more is known about phosphate transport.

Phosphate is clearly crucial to the ability of the mitochondria to synthesise ATP from ADP and there is good evidence for the presence of a *phosphate carrier* in the inner membrane. Phosphate uptake shows saturation kinetics: i.e. the rate of uptake approaches a plateau as the external concentration of phosphate is increased. Half-saturation occurs at a concentration of about 0.25 mol m^{-3} P_i and uptake is inhibited by reagents that bind to SH groups; by mercurial compounds such as mersalyl and by *N*-ethylmaleimide. These are diagnostic characters for the presence of carrier proteins in a membrane. Phosphate entry will occur down its electrochemical potential gradient. Entry does not alter the membrane potential so phosphate transport is electrically neutral; it must be accompanied by protons or be in exchange for hydroxyl ions. Either of these would dissipate the proton gradient, but maintain the electrical gradient.

Along with phosphate, the movement of adenine nucleotides between the mitochondrial matrix (the site of synthesis of ATP) and the cytoplasm is clearly central to the ability of the mitochondria to supply the cell with ATP for its various functions in the plant. Mitochondria export ATP and must take up ADP and P_i. The large electrical potential difference across the inner membrane is a powerful driving force for the export of ATP and the import of ADP, since ADP exists with one less negative charge than ATP (i.e. ATP^{4-} and ADP^{3-}). The negative electrical potential in the mitochondrial matrix relative to the cytoplasm drives out the ATP in exchange for import of ADP through an *adenylate carrier*. The carrier, which is inhibited by atractyloside, carboxyatractyloside and bongkrekic acid, has been isolated and characterised from animal, but not plant, mitochondria (see Prebble, 1988).

The synthesis of ATP depends upon the proton gradient generated from the oxidation of reduced coenzymes. These coenzymes are produced initially by the TCA cycle from the pyruvate supplied by glycolysis. Pyruvate, a small monocarboxylic acid ($^{-}OOC{-}C{=}OCH_3$), must enter the mitochondrial matrix for the process of oxidation to begin. Evidence that transport occurs through a specific *pyruvate carrier* comes from the fact that the uptake shows saturation kinetics (half-saturates at about $0.5\,mol\,m^{-3}$ pyruvate) and can be specifically inhibited by mersalyl and α-cyano-4-hydroxycinnamate. Pyruvate is presumed to enter by H^+-symport.

There are also porters for the di- and tricarboxylic acids of the TCA cycle and a specific carrier for oxaloacetate. Many of the intermediates of the TCA cycle are utilised outside the mitochondrion in synthetic reactions. If respiration is to continue when an intermediate such as α- ketoglutarate is withdrawn, another intermediate of the TCA cycle must be supplied from some other source within the cytoplasm. Consequently, there are carriers which facilitate the exchange of di- (e.g. succinic and malic acids) and tricarboxylic acids (citrate and isocitrate) across the inner membrane (Figure 4.4). Transport of dicarboxylic acids is mediated by a protein that exchanges acid for phosphate. Evidence for this *dicarboxylate carrier* arose from studies of the inhibitor *N*-ethylmaleimide, which inhibits phosphate/H^+ symport, but not phosphate transport in exchange for a dicarboxylic acid: the dicarboxylate porter is inhibited by *n*- butylmalonate. There is also evidence for dicarboxylate/H^+ symport (see Hanson, 1985). The *tricarboxylate porter* facilitates the exchange of malate for isocitrate or citrate. Citrate and isocitrate can also enter via symport with protons. Unlike the situation occurring in animal mitochondria, oxaloacetate crosses the inner

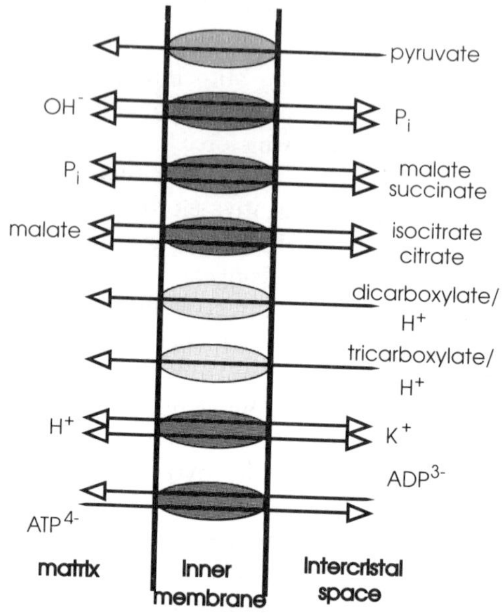

Figure 4.4 A diagrammatic representation of porters in the inner mitochondrial membrane. Pyruvate from the TCA cycle enters by proton symport. Phosphate either enters in symport with protons or exchanges with hydroxyl ions via the phosphate carrier. Phosphate can then exchange with the dicarboxylic acids, malate and succinate via the dicarboxylate carrier. The tricarboxylate carrier exchanges malate for isocitrate or citrate. Both di- and tricarboxylates can also enter in symport with protons. Finally, two antiporters are shown, one is for potassium and the other is the adenylate carrier.

membrane rapidly and apparently independently of fluxes of malate, although it is not finally proven whether or not there are two separate carriers for malate and oxaloacetate. Glycine, an important element in the traffic between mitochondria and peroxisomes appears to enter the mitochondrion by diffusion.

4.2 Chloroplasts

Within a photosynthetically active cell (tissue composed of cells containing chloroplasts is known as *chlorenchyma*), the chloroplasts follow the vacuole as the second largest compartment of the protoplast, occupying about 12–13% of the volume (Table 1.2). The chloroplasts, however, have a much higher concentration of enzymes than the vacuoles. They are, like the mitochondria, a major site of metabolism within cells and there are sizeable fluxes of soluble metabolites into and out of chloroplasts.

The chloroplast boundary is composed of *two* membranes making up an envelope which encloses the *stroma* within which lies a further complex arrangement of membranes, the *thylakoids* (Figure 4.5). The space bounded by the thylakoid membranes is called the *loculus*, and solutes are transported across the thylakoids in and out of this intramembrane space. The thylakoids contain the integral proteins that catalyse the light reactions of photosynthesis and the photosynthetic electron transport chain. The outer membrane of the chloroplast envelope is highly permeable due to the presence of porins (cf. the outer mitochondrial membrane). The inner membrane of the envelope has selective permeability, however, and is the site of carrier proteins that effect metabolite exchange between chloroplast and cytoplasm. The chloroplast is thus a more complex structure than the mitochondrion in that the membrane system involved in electron transport (the thylakoids) is itself contained within two membranes, one of which has selective permeability.

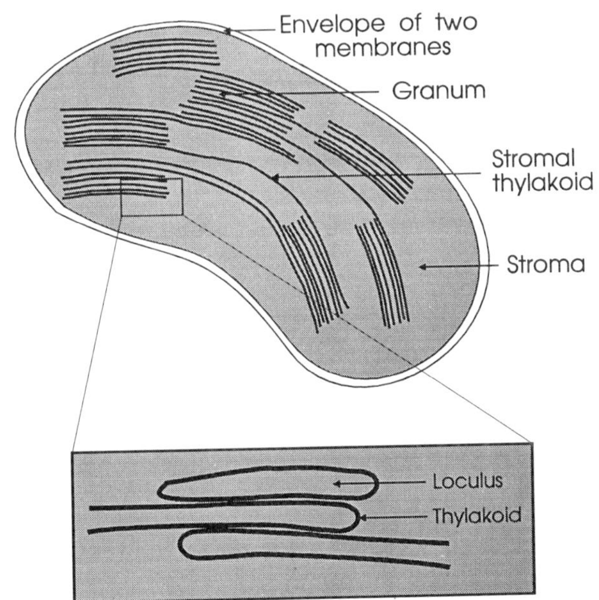

Figure 4.5 A diagrammatic representation of the structure of a chloroplast. Note that the inner membrane system is not connected to the envelope.

Plants in which the first products of carbon assimilation are the 3-carbon compounds of the Calvin cycle (hence they are known as C_3 plants) generally have chloroplasts of similar appearance throughout their chlorenchymatous cells. Carbon dioxide enters the chloroplasts and is fixed in a reaction catalysed by the enzyme ribulose bisphosphate carboxylase-oxygenase (Rubisco), a component of the Calvin cycle, producing 3-phosphoglycerate ($^{-}OOC–CHOH–CH_2OPO_3^{2-}$). The Calvin cycle is the only mechanism known to bring about the net fixation of carbon with regeneration of the primary acceptor. The fixed carbon is exported as a 3-carbon compound. This process of carbon fixation requires energy, which is derived from light. The currency of energy transfer remains ATP and the thylakoids are the site of an ATP synthetase. The major solute fluxes are, therefore, those associated with the generation of ATP by the ATP synthetase, the import of carbon dioxide and the export of fixed carbon. There are additional metabolite transport systems that allow the *net*

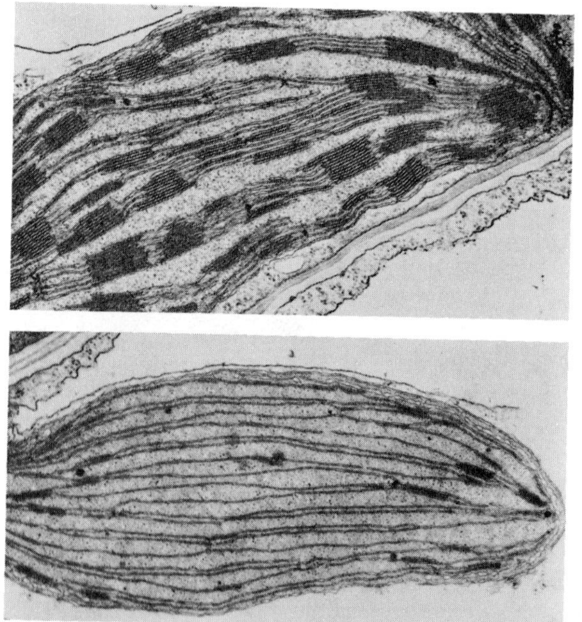

Figure 4.6 Mesophyll (upper, with grana) and bundle sheath (lower, agranal) chloroplasts in cells from a young maize leaf. The micrograph was kindly provided by Professor R.M. Leech of the University of York. Magnification × 18 900.

transport of ATP and reduced coenzymes from the stroma to the cytoplasm, the synthesis and export of amino acids (synthesis takes place within the chloroplasts, the site of many of the amino transferase group of enzymes) and the maintenance of an osmotic balance between cytoplasm and chloroplast.

Although C_3 plants possess chloroplasts which are fundamentally similar in appearance throughout their chlorenchyma, there is another group of plants in which chloroplasts differ in appearance between the mesophyll and the bundle sheath (Figure 4.6). In these species, the initial products of carbon fixation are 4-carbon compounds (hence C_4 plants). Carbon dioxide is first fixed in the mesophyll cytoplasm in a reaction catalysed by the enzyme phosphoenolpyruvate carboxylase to produce oxaloacetate ($^-OOC{=}O{-}CH_2{-}COO^-$). Subsequently, oxaloacetate is converted to malate or aspartate in the mesophyll chloroplasts, exported to the cytoplasm and then transported to the cells of the bundle sheath. Here, in the bundle-sheath chloroplasts, malate or aspartate enters the stroma, and is decarboxylated (Figure 4.7). The carbon dioxide released is refixed by the Calvin cycle. In these C_4 plants there are, therefore, additional obligate fluxes (of malate, aspartate and oxaloacetate) over those required in C_3 plants.

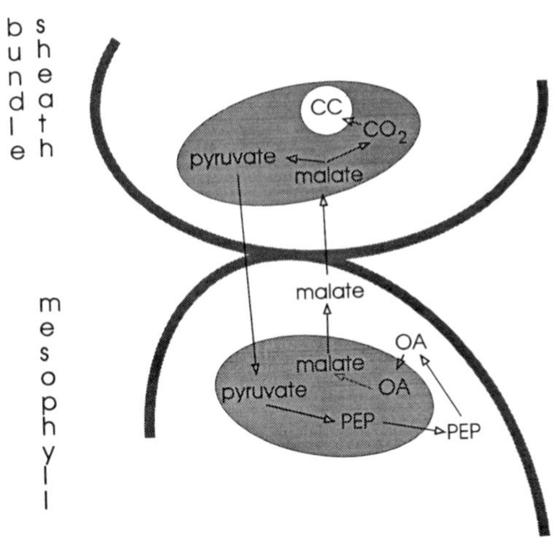

Figure 4.7 Carbon flows in C_4 photosynthesis. The diagram shows the interchange of metabolites between bundle sheath and mesophyll chloroplasts in a plant of the NADP-malic enzyme type. CC, Calvin cycle; OA, oxaloacetate; PEP, phosphoenolpyruvate.

The techniques that can be used to study solute fluxes across the chloroplast membranes are similar to those used with mitochondria. Since the chloroplasts are larger than mitochondria it is in many ways easier to study transport in the chloroplasts; volume changes can be calibrated using light microscopy, and the centrifugation required for measuring the accumulation and loss of radionuclides is simpler. However, chloroplasts are large and relatively fragile and cannot always be obtained in good condition by simple homogenisation of tissue; they are often prepared from protoplasts (cf. Figure 2.6). Isolation in non-aqueous media and centrifugation through silicone oils are all important techniques used in establishing the nature of the transport processes that occur (Sivak *et al.*, 1989).

Central to chloroplast function is the ability to synthesise the ATP necessary for the fixation of carbon dioxide and chloroplasts contain an F_0F_1 ATPase, located in the thylakoids. This enzyme is, like the mitochondrial enzyme, an ATP synthetase: it uses a proton gradient generated by the photosystems and the electron transport chain present in the thylakoid membranes to synthesise ATP. During this process protons are translocated into the loculus, generating a pmf across the thylakoid membrane with a raised pH and negative membrane potential in the stroma. In the light the pH in the stroma is around 8.0 while that in the loculus is close to 4.0. Membrane potentials are small, however, suggesting that they are dissipated by counter-ion fluxes of an anion or divalent cation (see Hedrich and Schroeder, 1989). There is evidence to suggest that Mg^{2+} is important in this respect, moving between the loculus to the stroma where its concentration rises by up to 5 mol m^{-3} in the light, activating enzymes of the Calvin cycle (see Haliwell, 1984). There is, then, a pronounced contrast between the chloroplast and mitochondrion. In the chloroplast, the free energy gradient of protons derives mostly from a gradient in chemical concentration (a pH difference of 4 units or 10 000-fold) and very little membrane potential, while in the mitochondria the membrane potential is much more important than the pH gradient.

The ATP synthesised in the chloroplasts is released within the stroma and used there during carbon fixation. Carbon enters the chloroplast as carbon dioxide. Although the carbon dioxide present in the cytoplasm is in equilibrium with carbonate and bicarbonate anions, the current consensus is that carbon dioxide itself crosses the envelope; being a small molecule it is freely permeable through the membranes. On either side of the envelope an equilibrium will be established between free carbon dioxide and the carbonate and bicarbonate anions. The bicarbonate anion appears to permeate only slowly (see Prebble, 1988). Once in the stroma, carbon dioxide is fixed by the Calvin cycle, generating 3-carbon compounds (triose phosphate) for export.

As far as the plant as a whole is concerned, it is the fact that carbon fixed in photosynthesis is exported from the chloroplast for use elsewhere that is crucial. Carbon is exported as triose *phosphate*: sucrose for long distance transport is synthesised not in the chloroplast, but in the cytoplasm. Since the export of fixed carbon as triose phosphate carries with it the export of phosphorus, there must be a mechanism for the return of P_i if the Calvin cycle is to continue. This is effected by an antiporter, that facilitates the exchange of triose phosphate (3-phosphoglycerate, dihydroxy acetone phosphate or 3-phosphoglyceraldehyde) for inorganic phosphate (Figure 4.8; see Heldt and Flügge, 1986; Flügge and Heldt, 1991). The porter, known as the *phosphate translocator*, has a molecular mass in its functional state of about 62 kDa made up of two monomers of about 30 kDa: it is located in the inner membrane of the envelope. There are two distinct functions fulfilled by this porter. First, it allows the export of fixed carbon as dihydroxyacetone phosphate or 3-phosphoglyceraldehyde in exchange for

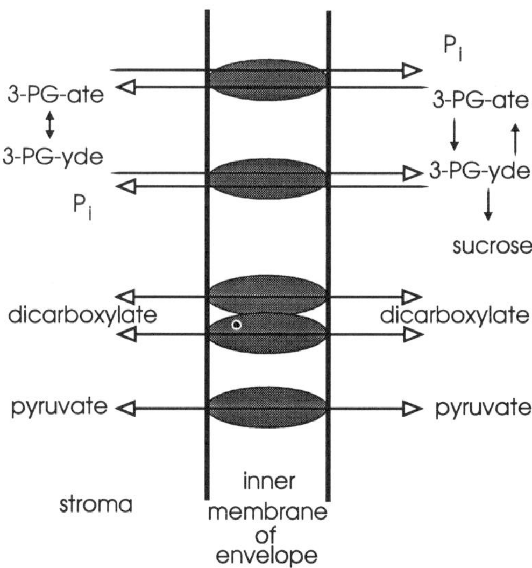

Figure 4.8 Carriers in the inner membrane of the chloroplast envelope. The phosphate translocator allows the exchange of triose phosphate (3-phosphoglycerate, 3-PG-ate, or 3-phosphoglyceraldehyde, 3-PG-yde) with phosphate. Sucrose may be synthesised from 3-phosphoglyceraldehyde in the cytoplasm or the 3-phosphoglyceraldehyde can be converted back to 3-phosphoglycerate, allowing the production of ATP, NADH and NADPH in the cytoplasm. Porters for dicarboxylates and pyruvate are also shown.

phosphate. This exchange maintains the stromal phosphate concentration and the exported triose phosphate is available for the synthesis of sucrose in the cytoplasm. Second, it effects the exchange of 3-phosphoglycerate across the envelope, thus allowing the net export of reducing power (NADPH) and ATP from the chloroplast to the cytoplasm (see Figure 4.8).

The exchange of triosephosphate across the envelope is one of the major fluxes associated with photosynthesis in C_3 plants. In C_4 plants (Figure 4.7), however, there must also be the exchange of oxaloacetate and malate across the chloroplast envelope. Specific translocators for dicarboxylic acids have, not surprisingly, been discovered in the chloroplast inner membrane (they are also present in the chloroplasts from C_3 plants). Dicarboxylate transport appears to involve two separate carriers with overlapping specificities (Figure 4.8; Flügge and Heldt, 1991; Prebble, 1988). A separate carrier effects the uptake of pyruvate (see Heldt and Flügge, 1986).

Chloroplasts also exchange metabolites with mitochondria and peroxisomes during the process of photorespiration (Figure 4.9). Glycollate export from the chloroplast is carrier mediated and coupled to glycerate uptake (see Flügge and Heldt, 1991). Phenotypic differences in photorespiration have been shown to be the consequence of mutations of carrier proteins. A particular mutant of *Arabidopsis thaliana* requires a

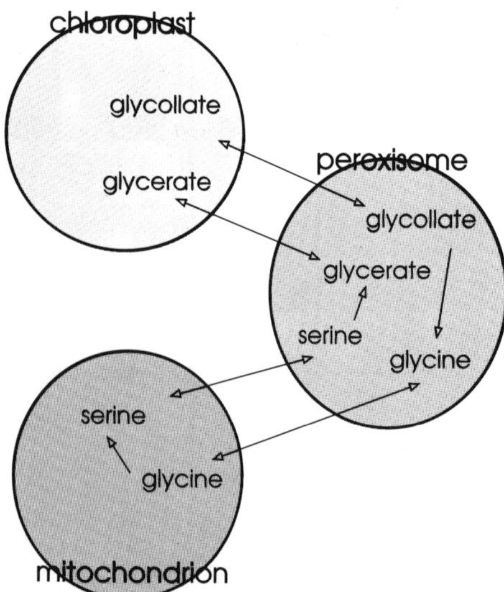

Figure 4.9 Carbon flows in photorespiration. The diagram shows the movement of metabolites between chloroplast, peroxisome and mitochondrion in a typical C_3 plant.

high ratio of carbon dioxide to oxygen (1% CO_2, 99% air) for survival (normal air is 0.03% CO_2, 20% O_2). This mutant, which is lethal, lacks a dicarboxylate carrier (Somerville and Ogren, 1983).

Metabolic activity in the chloroplast is characterised by the movement of solutes: fluxes across the thylakoid membranes and across the envelope. Since these membranes are semi-permeable, the solute fluxes will be accompanied by flows of water. In the light, chloroplasts do contract (see Nobel, 1984), but it is clearly difficult to try to integrate the net effect of the various fluxes that are taking place. Furthermore, there is little specific information about the regulation of transport of inorganic ions across chloroplast membranes, although it is clear that chloroplasts are able to regulate their internal ion concentration. For example, the chloride concentrations inside the chloroplast remain virtually constant in plants subjected to increasing external salinity, whereas the cellular chloride concentration rises dramatically (Flowers, 1984). Voltage-dependent chloride-selective channels have been identified using the patch-clamp technique in the thylakoid membranes of *Peperomia metallica* (see Hedrich and Schroeder, 1989).

4.3 Cytoplasm

In the previous section an important feature of solute transport was the exchange of metabolites across membranes. These solutes are presumed to be freely soluble within the aqueous phase of the cytoplasm, although they may bind to various surfaces. In other cases, metabolites are packaged within membranes for transport. This is generally the case for materials required for the synthesis of cellular components lying outside the plasma membrane, chiefly on the cell walls. In some cases the materials transported in these vesicles may no longer be soluble, but in others they are presumed to be so. There is also evidence for the uptake of solutes in vesicles.

4.3.1 *Vesicular transport systems*

Perhaps the best studied of vesicular systems is that originating from the *Golgi* apparatus. The Golgi bodies or *dictyosomes* are stacks of membrane-bound disc-shaped sacs, known as *cisternae* (see Figure 2.1). Vesicles are formed from the cisternae which in turn are derived from the endoplasmic reticulum. The vesicles eventually fuse with the plasma membrane in a process of *exocytosis*, releasing the contents of the vesicle to the outside of the cell membrane. The Golgi are, then, part of a flow of membranes within the cell as well a medium of transport.

There is a variety of evidence to connect the Golgi vesicles with the transport of solutes to extracellular spaces. Some of this is circumstantial, such as that from electron microscopy. For example, the Golgi are more prominent and the vesicles larger in cells which have a secretory function than in those that are not active in this respect; the Golgi vesicles are particularly well developed in cells from root caps that are secreting mucilage; the tips of pollen tubes, which are a good example of directed growth, are remarkable for the predominance of vesicles at the growing point; in the alga *Pleurochrysis*, polysaccharide scales destined for the outside of the plasma membrane can be seen within the Golgi vesicles on the way to their final destination (see Hall *et al.*, 1982). There is also autoradiographic evidence obtained by electron microscopy. In these experiments a radioactive precursor is fed for a limited time (the pulse) to a plant and the localisation of the isotope followed with time from the end of the pulse, during which time the radioactive precursor is replaced with its non-radioactive form (the chase). These time series demonstrate the movement of label from Golgi vesicles to an external position. Little is known, however, of how the secretions are regulated or directed to their final destination, although the latter is likely to involve internal microtubules (see below).

Vesicles are also involved in the uptake of materials by cells, more so in animals than plants. The process is known as *endocytosis*. Endocytosis is the uptake of extracellular substances by invagination of the plasma membrane. In animals, where endocytosis is a well established method of solute uptake, the solutes are enclosed within a special class of vesicles known as *coated vesicles*; coated since the cytoplasmic face of the membrane bears a protein coat visible under the electron microscope. One of the best known of these proteins is *clathrin*, a protein with a molecular mass of about 630 kDa (see Robinson and Depta, 1988). In some cases the vesicle formed simply engulfs fluid (*fluid phase endocytosis*) and the solutes dissolved in that fluid: in other cases specific molecules in the medium bind to receptors on the membrane before being engulfed. The latter takes place in the region of *coated pits* (see Robinson and Hillmer, 1990) where protein coats are visible on parts of the plasma membrane. Rates of fluid phase endocytosis can be very high in some animal systems, representing the conversion of 3% of the plasma membrane to vesicles per minute. Whether or not coated pits are involved in this form of endocytosis is controversial.

In the case of plant cells, two factors complicate the occurrence of endocytosis: the presence of the cell wall and of turgor pressure. The cell wall interferes with the movement of large molecules up to the plasma membrane and the membrane will have to invaginate against the pressure arising from cell turgor. Nevertheless, that endocytosis does occur in plants can be argued from a theoretical point of view: there must be some means of removing additional membrane material deposited in the plasma membrane during exocytosis. Furthermore, coated pits are a feature of the plant plasma membrane. It appears from calculations that endocytosis is only possible in plants where the endocytotic vesicles are small (100 nm, the size of coated pits) and where the cells have relatively low turgor (see Robinson and Hillmer, 1990). Experimental evidence consistent with the occurrence of endocytosis in plants comes from the use of a fluorescent probe, Lucifer Yellow CH. This compound has a molecular mass of 457 Da, is negatively charged, has a very low toxicity and does not cross membranes. If injected into plant cells it does not diffuse out (it can, however, be transported from cell to cell via the plasmodesmata, see below). The fact that the dye is not taken up by *isolated* vacuoles but does enter cells and vacuoles when, for example, roots are incubated in a solution of Lucifer Yellow, can be explained by the occurrence of fluid phase endocytosis. Other evidence comes from the uptake of electron dense particles of ferritin (an iron-containing protein) by protoplasts and vesicles and the observation that electron dense deposits can be seen in vesicles of cells exposed for a few hours to an aqueous solution of salts such as lanthanum nitrate (lanthanum is electron dense: see Robinson and Depta, 1988). There is also some evidence for receptor-mediated uptake by plant cell membranes. Vesicular uptake appears to be on a pathway to the vacuoles—a similar pathway to that in animals, where the contents of coated vesicles are delivered to lysosomes. The similarities between the plant vacuole and the animal lysosome have often been remarked upon (Hall *et al.*, 1982).

Endocytosis, exocytosis, the Golgi, the endoplasmic reticulum and the vacuoles are part of a system of membrane turnover operating in cells. In some instances, the turnover of the plasma membrane can be extremely rapid; just 10 minutes in secretory cells and only 3 hours in growing epidermal cells, as judged from stereological measurements (Steer, 1988). This dynamism of the plasma membrane casts a new light on the permanence of ion pumps and channels embedded in it. It is not known whether these proteins are continually reinserted into the plasma membrane or whether only parts of the plasma membrane turn over rapidly.

4.3.2 *Microtubules*

These are tubular structures present in nearly all eukaryotic cells. They have an average exterior diameter of about 24 nm and are assembled from subunits of *tubulin*. They can change their length through assembly or disassembly of their subunits and are sensitive to specific chemicals such as colchicine. Tubulin is a protein composed of two (α and β) subunits of approximately the same molecular mass, about 55 kDa, as determined from electrophoresis. The amino acid sequence of the subunits has been determined and suggests a molecular mass of 50 kDa (see Dustin, 1984). The microtubules are most commonly composed of 13 rows of subunits or protofilaments. Each protofilament is composed of alternating α and β

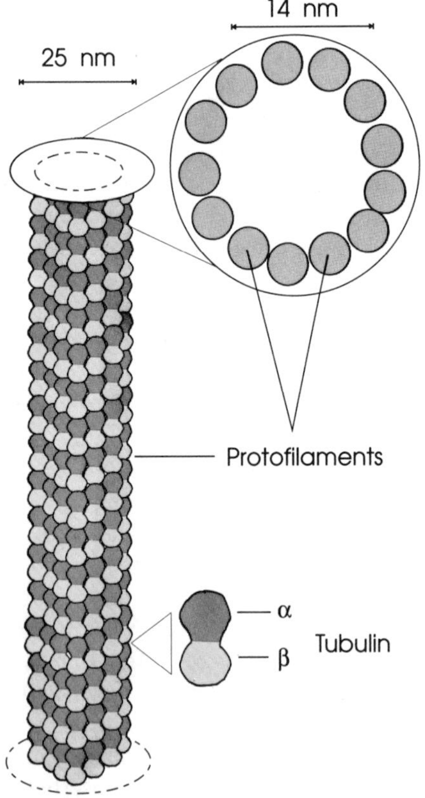

Figure 4.10 A diagrammatic representation of a microtubule composed of protofilaments of α and β tubulin.

subunits (Figure 4.10). These subunits can assemble *in vitro*, the process requiring the hydrolysis of guanosine triphosphate (GTP).

Microtubules, which during interphase radiate from the nucleus towards the periphery of the cell (see Lloyd, 1987), are involved in intracellular movement that is sudden and orientated, *saltatory movements*. A well known example is the movement of chromosomes during cell division (see Baskin and Cande, 1990). Such directed movements may have a role in solute transport, for example in the transfer of mitochondria to a specific site within a cell where ATP is required (see Douce and Neuberger, 1989).

4.3.3 *Cytoplasmic streaming and intracellular diffusion*

Although electron micrographs present a static picture of the cytoplasm within cells, cellular contents are in continuous and sometimes rapid motion. Brownian motion of small particles can be seen as well as a general movement of the cytoplasm. Cytoplasmic movements are of two types; first, the streaming associated with changes in cell form such as seen in the movement of slime moulds and second the circulation of the cellular contents seen in most plant cells (Kamiya, 1981). The latter, *cytoplasmic streaming* or *cyclosis*, is important in the movement of solutes within a cell, as streaming can lead to more rapid transport than by diffusion alone. The velocity of cytoplasmic streaming is commonly about 5 $\mu m\ s^{-1}$. Streaming can also reduce the width of the boundary layer on the inside of a membrane, thus increasing rates of diffusion across membranes, where the rate is limited by the thickness of the boundary layer.

Cytoplasmic streaming, which has been extensively investigated in giant cells of the characean algae *Chara* and *Nitella*, is thought to be brought about by the movement of myosin attached to the endoplasmic reticulum and to organelles along actin filaments within the cytoplasm (Kuroda, 1990). Both actin and myosin have been identified within plant cells and streaming is dependent upon the presence of ATP and Mg^{2+} (both about 1 $mol\ m^{-3}$) and Ca^{2+} (less then 0.1 $mmol\ m^{-3}$). Streaming requires a force of 0.1–0.2 $N m^{-2}$ (or 0.1–0.2 Pa) and is inhibited by cytochalasin. It is still not known whether microtubules play a role in cytoplasmic streaming or simply define the route along which streaming occurs.

4.4 Intercellular transport

Solutes move not only within cells, but between neighbouring cells and between more distant parts of the plant. For example, photosynthate must be transported from the leaves to the roots during normal growth and to developing seeds after flowering. This transport takes place in the cells of the phloem and xylem (see chapters 6 and 7), tissues specialised for long distance transport. However, xylem and phloem do not adjoin every cell: local transport must take place between cells, at least over distances of a few to tens of cells.

4.4.1 *Symplast and apoplast*

Water and dissolved solutes move from root to shoot in the xylem, in conduits of dead cells. These cells form the most obvious feature of a pathway in which solute transport can occur external to the living parts of the cell and the plasma membrane. This phase has been given the name *apoplast*: it consists of cell walls, intercellular spaces and the lumina of non-living cells such as the xylem vessels. Transport, however, does not only occur in the apoplast. There is also a remarkable pathway within the living cells, for plant cells are interconnected through narrow cytoplasmic channels. This has lead to the development of the concept of a *symplast*—all the connected cytoplasm within a plant (synonym symplasm). A solute in a particular protoplast need not have entered that particular cell, but may have reached its present location after entering the symplast at any point within the plant. Transport can occur in either phase, through the symplast or the apoplast (Figure 5.6), and many factors affect the relative importance of each pathway.

4.4.2 *Plasmodesmata*

Cytoplasmic connections between neighbouring cells of plants were suggested before the turn of the century and the German botanist Strasburger used the term 'plasmodemen' to describe them in 1901. The term *plasmodesma* (plural, *plasmodesmata*) remains in use to describe cytoplasmic connections between cells. They have a diameter of about 60 nm and are visible under the electron microscope (Figure 4.11). Plasmodesmata have been described from all the major groups of plants. Although it is accepted that they arise during cell division as persistent protoplasmic threads, plasmodesmata do also occur in cell walls which have not come directly from a cell division and between the cells of hosts and parasites (see Robards, 1975; Robards and Lucas, 1990). This suggests that they can form and pass through an existing wall. Plasmodesmatal frequency shows a great range, between about 0.1 to 10 per μm^2 (Robards and Lucas, 1990) with the

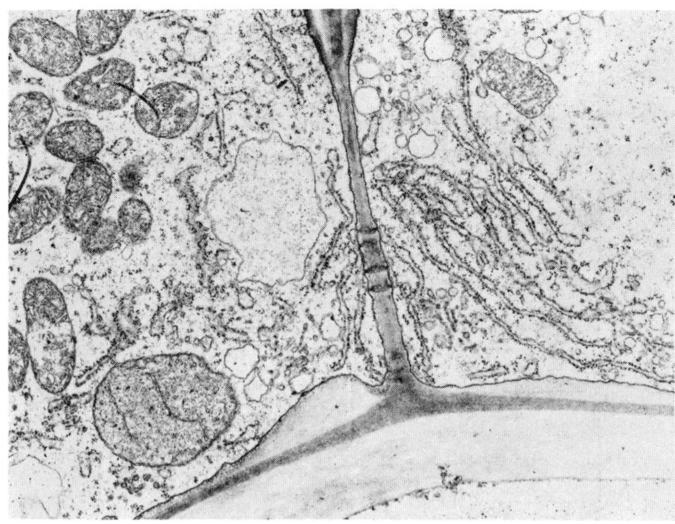

Figure 4.11 An electron micrograph of adjacent parenchymatous cells from the xylem of a mature maize root. Plasmodesmata are clearly visible crossing the cell wall between two cells. Rough endoplasmic reticulum can be seen in both cells and appears to pass through the plasmodesmata. The cell on the left contains mitochondria, a proplastid and a microbody. Magnification × 10 175.

most common number being between 1 and 10 per μm^2. Thus even a small meristematic cell (a cube of 10 μm side; see section 1.1) would have between 600 and 6000 plasmodesmata.

It has proved difficult to describe the detailed structure of plasmodesmata, as they are so small. Electron microscopy has shown that the plasma membrane lines the plasmodesma, which also contains a central structure, the *desmotubule* (Figure 4.12). This is thought to consist of a tightly rolled cylinder of membrane with virtually no central lumen. The only lumen through a plasmodesma is the annular channel between the desmotubule and the lining plasma membrane, known as the *cytoplasmic sleeve*. This annulus is not a clear channel, however, and contains an array of particles, the exact number of which either varies between tissues or is ill-defined. At present the consensus is that there are nine particles, although in one species the number is six. The plasmodesmata are commonly restricted at each end, at the neck (see Figure 4.12). Such restrictions may play a role in regulating the passage of materials through the symplast, the neck constriction being seen as part of a sphincter able to regulate the opening and closing of the plasmodesmata.

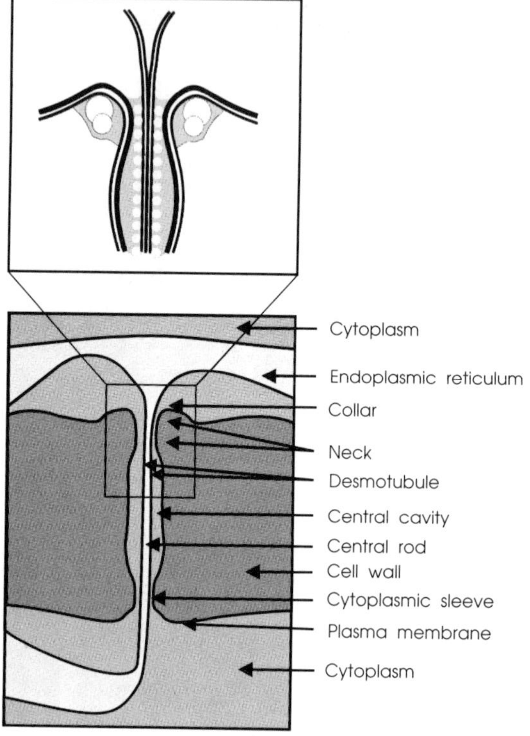

Figure 4.12 A diagram of a longitudinal section through a plasmodesma, with an enlargement of the neck region. Redrawn from Robards and Lucas (1990).

The function of the plasmodesmata has long been assumed to be in the transport of substances of small molecular mass and of electrical currents. This is supported by circumstantial evidence that the frequency of the plasmodesmata is related to the extent to which a cell is involved in transporting materials. Additional evidence comes from autoradiographs, which have shown chloride ions to be located within the plasmodesmata of tissues transporting these ions, and the low electrical resistance occurring between the giant internodal cells of algae such as *Nitella*. Recently, analysis of the movement of fluorescent dyes through the plasmodesmata (e.g. Terry and Robards, 1987) has suggested that transport occurs in the cytoplasmic sleeve or annulus through discrete channels existing between the nine particles associated with the desmotubule. Dimensions quoted for the various components of the plasmodesma are 5.6 nm for the sleeve and 4.5–5 nm for the particles associated with the desmotubule. The gaps between the particles are then between 1.5 and 3.5 nm (quite similar to the limiting pore size of plant cell walls, see section 5.1.6), giving a cross-sectional area of between 2.6 and 3.6 nm^2 for the spaces in which solutes can travel. This is similar to the cross-sectional area of the pores in animal gap junctions (3.1 nm^2; Robards and Lucas, 1990). There is another similarity between the plasmodesmata and the gap junctions of animal cells (transmembrane protein channels produced from six polypeptide chains containing a single aqueous channel of about 2 nm diameter); an homologous protein appears to be present in the two structures (see Meiners *et al.*, 1988).

4.5 Glands

Vesicular transport systems in plants are particularly associated with movement of substances from symplast to apoplast: from within to without the plasma membrane. Such transport occurs in all cells during the deposition of the cell wall. There are, however, cells and groups of cells particularly specialised in this form of transport, i.e. the transport of salts and sugars outwards across the plasma membrane. In this section we will describe the operation of glands that secrete sugars: coverage of salt glands is left to chapter 6, where the context in which they operate is explained better.

The various epidermal glands found in plants are formed from specialised cells termed *trichomes*. These are unicellular and multicellular appendages of the epidermis (Fahn, 1988). In some carnivorous plants these glands secrete mucilage (polysaccharides) and digestive products across the plasma membrane in vesicles derived from the Golgi apparatus, while oils are produced from glands in plants such as thyme (*Thymus* spp.). Yet other glands (*colleters*) produce the sticky mixtures of terpenes and mucilages, commonly found on bud scales (for example in the horse-chestnut, *Aesculus hippocastanum*). The stinging nettle (*Urtica dioica*) has hairs, impregnated with silica at their tips, which can penetrate the skin of animals and deliver the contents of the epidermal cell. Salt glands and nectaries secrete water soluble substances. The secreted materials may either cross the plasma membrane directly (*eccrine secretion*) or in vesicles (*granulocrine secretion*).

4.5.1 *Nectaries*

Nectaries are specialised tissues in the plant which secrete a sugary solution, *nectar*: they may be floral or extrafloral. Nectaries consist of groups of specialised epidermal cells overlying small parenchymal cells with thin walls and dense cytoplasmic contents. Nectar may be secreted by the epidermal or the parenchymal cells. Whereas salt glands tend to be isolated (even if consisting of 40 cells), nectar secreting cells are gathered together in groups often large enough to be visible to the naked eye (see Fahn, 1988).

The solution that is produced in nectaries consists chiefly of a mixture of glucose, fructose and sucrose (Table 4.1), but it also contains amino acids, organic acids and salts (see Fahn, 1979). Secretion can take place from a great range of structures (Fahn, 1988): from the unicellular trichomes such as those found in the nectaries of *Lonicera* (honeysuckle); from multicellular trichomes as in the nectaries of *Abutilon* (abutilon); from nectariferous tissue that underlies modified stomata (e.g. *Colchicum*, autumn crocus) and from the septal nectaries in the ovaries of monocotyledonous plants (e.g. *Muscaria*, the grape hyacinth). Whatever the nature of the nectaries, however, they are usually to be found close to the phloem (chapter 7) and there has been considerable interest in whether the pathway of transport to the secreting cells is apoplastic or symplastic.

Nectariferous tissue is generally in apoplastic continuity with the cells of the leaf and of the phloem, although there may be a barrier between certain of the secreting cells and the remaining tissue. Thus in the multicellular glands of *Abutilon*, the cell at the base of the trichome has a barrier to apoplastic transport (see Findlay, 1988). There is, on the other hand, symplastic continuity through plasmodesmata between the phloem and the cells of the trichome. The symplast looks the likely pathway of transport of sugars in this case.

Table 4.1. Concentration of glucose, fructose and sucrose in nectar from three plant species. The data are taken from original papers quoted in Findlay (1978) and Fahn (1979)

	Concentration ($mol\ m^{-3}$) of various solutes in		
Solute	*Ricinus*	*Vigna*	*Musa*
Glucose	1650	1400	444
Fructose	1639	1400	444
Sucrose	819	792	526

The cells of some nectaries show extensive wall protruberances, a characteristic of *transfer cells*, and secretion is presumed to take place through the plasma membrane. Elaboration of the cell wall allows an increase in the surface area of the plasma membrane. In other tissues (e.g. the banana, *Musa*) there is clear evidence from electron micrographs of intense Golgi activity and abundant endoplasmic reticulum. Whether the Golgi activity is associated with the secretion of nectar or the elaboration of the cell wall cannot be easily determined, because of the difficulty of localising small organic molecules using the electron microscope (see section 1.3.1.). Where there is an abundance of endoplasmic reticulum this may be related to membrane turnover in the cell (cf. section 4.3.1).

The process of secretion is poorly understood in biochemical terms. The dense cytoplasm and presence of mitochondria suggests that the transport process is an active one. Whether transport involves sucrose alone or sucrose, glucose and fructose is not clear. The nectaries are certainly sites of high invertase activity (invertase catalyses the hydrolysis of sucrose to glucose and fructose) and it is possible that glucose and fructose are produced extracellularly following sucrose transport across the plasma membrane. The nature of sugar transport and phloem unloading is discussed in more detail in chapter 7. An interesting possibility is the presence of a sugar–proton cotransport system.

CHAPTER FIVE

ION UPTAKE BY PLANT ROOTS

The uptake of ions by plant roots involves a series of events. First of all, the root must intercept the ions. This is a matter both of ions moving towards roots and of roots moving towards ions. The overall process is commonly aided by fungal associations that increase the effective radius of the roots. Ions must then be taken up by the root and transported radially inwards by one or more of a number of available pathways. Finally, those ions destined for the shoots must enter the tissues that have evolved for long distance, high speed transport. To do this the ions have to cross a barrier whose functions are to impose selectivity on the constituents of the external solution entering the plant and to prevent the concentrated solution in the transport system from leaking out.

5.1 Movement of nutrients towards the root plasma membrane surface

5.1.1 *Boundary layers*

Roots are surrounded by a *boundary layer* or *unstirred layer* of solution (see Figure 2.7), which usually limits the rate of nutrient absorption. It is analagous to the boundary layer of air that surrounds leaves. The thickness of this layer is smallest where roots are growing in rapidly moving solution and increases in poorly stirred solution and in the soil. Solutes of nutritional importance are accumulated through the plasma membrane of cells near the surface of the root and this has the effect of depleting the concentration of those nutrients in its immediate vicinity. This local depletion generates a concentration gradient down which the solutes must *diffuse* from the bulk (soil) solution. At concentrations typical of soil nutrients, the diffusive flux supplying the root surface is lower per unit of membrane surface than the flux across the membrane into the plant. The result is that the average concentration presented to the membrane is lower than that in the solution. For example, with an unstirred layer of 100 μm and a nitrate concentration of 0.1 $mol\,m^{-3}$ in the bulk solution, the presence of the unstirred layer will reduce the nitrate concentration at the root surface by two orders of magnitude (see Clarkson, 1988).

Other than diffusion, the only physical agency tending to replace the nutrients removed at the root surface is *advection*, a mass flow of soil solution towards the root to replace the water lost with the transpiration stream. This is the main pathway by which water usually reaches the root surface (see Taiz and Zeiger, 1991). Advection is a major factor in nutrient supply only at unusually high external concentrations. In all other situations, where the concentration of nutrients in the transpiration stream is higher than their concentration in the soil solution, nutrients are removed from the root surface at a rate greater than that at which mass flow could replace them.

5.1.2 *Exploration by roots*

As plant grows, so the depletion of nutrients from the surrounding soil intensifies and, because diffusion becomes even less effective the greater the distance (cf. section 2.4), supplies of nutrients to the root become increasingly limited. Continuous extension of the root system is therefore necessary to explore and exploit new volumes of soil solution. Since roots are capable of elongating at much greater rates than ions can diffuse through the soil, young roots are likely to be in higher nutrient concentrations than mature roots.

5.1.3 *Root hairs*

Root hairs develop as protuberances from the epidermis (Figure 5.1). Their role has been associated with the fact that they increase the surface area of the epidermis to a considerable extent. However, the fact that depletion zones develop rapidly around roots does not support the view that the absorptive surface area is itself a limitation to ion accumulation. Clarkson (1985) points out that the increase in effective root radius produced by the length of the root hairs is much more important than their surface area. This increased radius extends the epidermal surface outwards by a distance of similar magnitude to the thickness of boundary layers and may be particularly important in the movement of water to roots in drying soils. Root hairs also have a mechanical role in supporting the extending root (Clarkson, 1974) since considerable pressure may be needed for the apex to force a path through the soil particles.

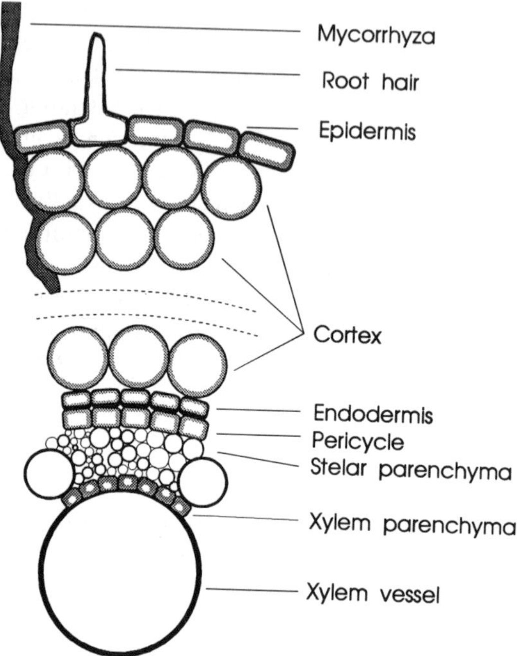

Figure 5.1 A stylised cross-section of a mature root showing the major zones, the epidermis, cortex and stele, and the principal cell types.

5.1.4 *Fungal symbionts*

The roots of most plants in the field are involved in symbiotic associations with fungi; the most common type are the *vesicular-arbuscular*. The interface between fungus and plant occurs at developments called *arbuscules* in the cortical cells. Extensive invaginations of the plasma membrane, analagous to those seen in transfer cells (section 4.5.1), provide a large surface area across which the currencies of the symbiotic relationship pass. Fixed carbon from the plant is exchanged (metaphorically) for nutrients which the fungus has acquired from the soil. The fungus extends from the plant root into the soil in the form of long filaments or *mycorrhizae* (literally *fungal roots*). Mycorrhizae also increase both the effective absorptive surface of the root and the volume of soil that it can access: the mycorrhizae penetrate very much greater distances into the soil than root hairs do. The importance of fungal associations is thought to be at its greatest for the absorption of phosphate ions, which have very poor mobility in soil and so do not diffuse easily towards the root (Clarkson, 1988). Increased uptake of phosphate by roots infected with mycorrhizae has been demonstrated to be due to the more efficient exploration of the soil, rather than to mobilisation of phosphate sources which were unavailable to plant roots (Sanders and Tinker, 1971).

5.1.5 *Ion exchange*

Soils, particularly clay minerals, have an exchange capacity for cations. This can provide a reservoir of nutrients which exists in equilibrium with the soil solution. If an ion is depleted from the soil solution it may be replaced from the reservoir to maintain the equilibrium between free and exchangeable ions. This equilibrium may be modified by the efflux of protons across the plasma membrane because protons may displace other ions from exchange sites. Nutrient supply and availability in clay soils is, as in the cases of slow-release fertilisers and chelated nutrients, a much more complicated matter than it is in simple solutions.

5.1.6 *The cell wall*

Any solute entering the root must move across, and possibly along, cell walls to gain access to the plasma membrane of the cells. Even the primary cell wall is extremely complex, but taking a simplistic look at some growing tissues, their polysaccharides can be fractionated into three major classes, cellulose, pectins and hemicelluloses. Cellulose is probably the most well known of these polysaccharides: it is 90% glucose, perhaps 100% glucose, and where other sugars or sugar residues are reported, these may be contaminants. Glucose units are joined by β-1,4 linkages, 200 to 14 000 units forming a molecule 1–7 μm long with a molecular mass of about 1000 kDa. Individual molecules line up to produce orientated aggregates termed *microfibrils* (see Preston, 1974), with dimensions of about 10×4 nm. The spaces between the fibrils, that is the effective pore sizes of cell walls, has been determined empirically. Carpita *et al.* (1979) used high concentrations of a range of molecules (such as polyethylene glycols and dextrans) whose molecular size was known and observed which molecules caused plasmolysis and which cytorrhysis: plasmolysis is the shrinkage of the protoplast away from the cell wall caused by the osmotic withdrawal of water and cytorrhysis is the shrinkage of the whole cell. For plasmolysis to occur, the solute must have crossed the cell wall in order to act across the plasma membrane resulting in the protoplast shrinking away from the wall.

In cytorrhysis the solute does not cross the cell wall and the whole cell collapses as water is withdrawn with the cell wall acting as the 'semi-permeable membrane'. The effective pore size in the walls of a range of species was found to be from 3.8 to 5.2 nm, similar to the spaces thought to be available for solutes to pass through plasmodesmata (cf. section 4.4.2). Pore size alone, however, is not likely to be a limitation for the uptake of most nutrients, but two other aspects of the cell wall are. The first is that the pathway between the microfibrils is very tortuous and diffusion is likely to be much slower than in free solution. The second is that cell walls have a large excess of net negative charge which will interact with different charged solutes in different ways. These charges are carried on the molecules that make up the pectins. Particularly important from this point of view, is polygalacturonic acid, a polymer containing α-D-1,4 linked galacturonosyl residues that turns the cell wall into an ion exchanger (a Donnan system) with important consequences for the concentration of particular ions in the neighbourhood of the plasma membrane (see section 5.5.1).

5.2 Ion uptake and the structure of the roots

A transverse section through a differentiated region of a root shows a concentric arrangement of cylinders of cells. Solutes and water must pass through or around these cell layers from epidermis to xylem vessel (Figure 5.1). A longitudinal section shows that all of these cells change with distance from the apex, as the root ages, in a complex interplay of real and physiological time. The developmental state of a cell which is a fixed distance behind the apex is a function both of the time that has elapsed since it was produced in a mitotic division in the apical meristem, and of all the factors, including climate and nutrition, that will have affected the plant during that time. This complex four-dimensional picture of the root is fortunately capable of reasonable simplification.

The *endodermis* has long been recognised as demarking a radial boundary: it is the inner border of the root cortex. Up to and beyond the endodermis, solutes are able to move in both apoplastic and symplastic pathways—the cell wall/extracellular spaces and cytoplasmic/plasmodesmatal continuites, respectively—as well as passing across membranes and cell walls in the so-called *transmembrane path*. Modifications of the cell wall at the endodermis, a cylinder of tightly adpressed cells without intercellular spaces, limit the pathways possible. The differentiation of the endodermis is characterised by the formation of

Casparian bands, in which substances impervious to water are deposited in all of the radial cell walls. Further, the plasma membrane is more tightly adpressed to the cell wall in the endodermis than most other tissues in the plant; if the endodermis is plasmolysed the plasma membrane is often seen to remain adhering to the wall at the site of the Casparian band. In its primary state (Figure 5.2) water and solutes can pass radially across the endodermis by traversing the plasma membranes of the outer and inner tangential walls, as well as crossing them within the symplast via plasmodesmata. In some species the endodermis is later shed with the cortex during secondary thickening. In these species, which are all dicotyledonous, a new protective layer, the *periderm*, develops from a cork cambium which arises in the root pericycle. The periderm then becomes the barrier to the radial diffusion of solutes. In other species, which do not undergo secondary thickening, further developments take place within the endodermis. A layer of suberin, impervious to water, is deposited between the plasma membrane and the entire surface of the tangential wall (Figure 5.2). At this (secondary) state, water and solutes cannot traverse the endodermis in any way other than in the symplast via the plasmodesmata. Further wall thickenings are laid down as the endodermis develops into the final (tertiary) state, but this has no further effect upon the pathways available to radial movement. The changes within the walls of the cells of the endodermis block the apoplastic pathway with considerable efficiency. This is shown clearly in micrographs which localise particles of lanthanum hydroxide that are both electron dense and membrane-impermeable; they are seen to accumulate up to, but not beyond, the Casparian band (see Clarkson, 1974).

The stoppage of the apoplastic pathway in the endodermis has two functions. One is to impose membrane selectivity on almost all uptake of solutes by the plant. Biological (membrane) selectivity in the xylem is discarded early in its development in favour of the high axial conductance needed to provide water to the transpiring shoot. The cells of the xylem die to form a series of pipes; the mature xylem vessels (see section 6.1). Without the endodermis a mass flow of solution would be possible from the outside to the xylem via the apoplast and hence to the shoot. The selectivity of this pathway would only be physical (mostly effective pore size) and chemical (mostly the exchange characteristics dictated by the fixed negative charges of the components of the cell walls). The plasma membrane of the cells of the endodermis becomes the barrier that all substances moving apoplastically have to cross to gain access to the conductive tissues of the stele. A small leakage, called the *bypass-flow*, occurs in young roots before the endo-

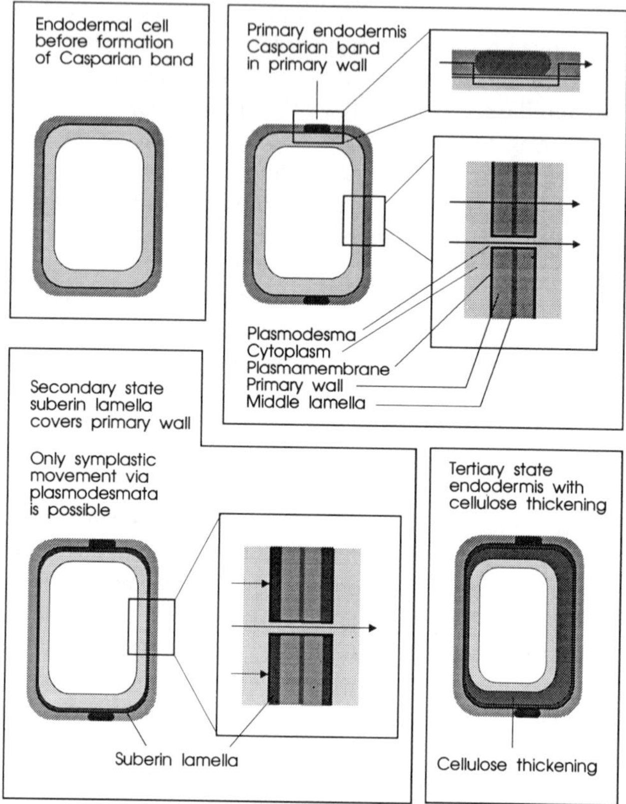

Figure 5.2 The stages in the development of the endodermis.

dermis has differentiated and in mature roots where the endodermis is disrupted by the emergence of lateral roots from primordia in the pericycle (the first cell layer within the endodermis; Figure 5.1). Of course, damage and disease will have major effects. Any gardener would recognise that the state of roots harvested from the soil after disease, various herbivores and the physical stresses of expansion and contraction of the soil have had their way, is far removed from the pristine condition of roots of hydroponically-grown seedlings. The importance of prevention of radial leakage from outside to xylem can be appreciated most clearly when high concentrations of potentially toxic ions are present externally, such as in saline soils (see chapter 6).

In the great majority of situations, plant roots are not exposed to large quantities of potentially toxic substances, and only a small proportion of the total water uptake is thought to cross the root in the apoplastic continuity (Boyer, 1985), so it would be surprising if prevention of inwards apoplastic leakage was the only reason for the radial differentiation we observe in the root. A second function of the endodermis is to prevent ions leaking *out* of the root. The chemical potential of most nutrient solutes in the xylem is higher than in the medium external to the root and often by two or more orders of magnitude. These solutes could diffuse out of the root if the cell wall continuity were not blocked.

The endodermis separates the root into two regions, the cortex-plus-epidermis in which solutes are acquired from the external medium, and the *stele*, in which they pass into the conductive tissues and from there to the rest of the plant. Recent studies with fluorescent dyes have highlighted the presence of another layer of cells that generally contains Casparian bands (88% of the families of angiosperms): the *hypodermis*. This is the outermost layer of cells in the cortex: a hypodermis which contains Casparian bands is termed an *exodermis* (Peterson, 1988). The exodermis, where it develops in older regions of the root, is likely to constrain to the epidermal cells the uptake of ions by restricting their access to the epidermal apoplast alone. If the differentiation of the mature root appears to impair the uptake of solutes, it must be remembered that for plants in the soil, nutrient depletion means that mature roots would have a limited part to play in nutrient uptake.

5.3 Phenomenology of ion uptake by roots

The plasma membrane of the epidermal and cortical cells of roots can mediate in the accumulation of nutrients into the symplastic continuity by the processes described in earlier chapters (entry down free energy gradients, and primary and secondary active transport). The epidermis, including its root hairs, is the outermost layer of the root and the first to come into contact with the (soil) solution (but we should not forget the mycorrhizae). Subsequently ions reach the cortex. The role of the cortex in ion accumulation is uncertain. The large surface area available for uptake into the symplast by summing the surface area of all the cells of the epidermis and cortex is often emphasised (see, for instance, Lüttge and Higinbotham, 1979). However, the extent of the involvement of the cortex in solute uptake into the symplast depends upon the solute and its

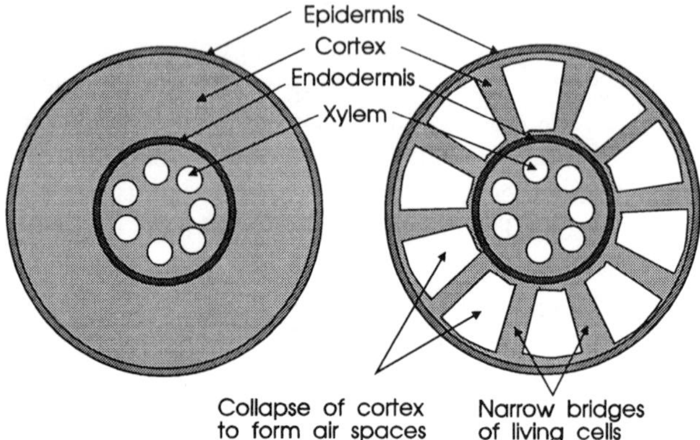

Figure 5.3 A diagram showing the consequences of the development of aerenchyma on the surface area available for the transport of ions across the cortex of a root.

concentration: the depletion zones around roots testify that in many cases the epidermis is capable of accumulating solutes faster than diffusion and mass flow can supply them. Most uptake of ions by plants in the field is thought to take place at the epidermis and its associated root hairs (Nye and Tinker, 1977). In the solution that penetrates beyond the root surface, where there is no exodermis, there may be few nutrients left to accumulate. A further observation that demonstrates that the surface area of the cortex is not necessary for ion uptake is that the disruption of the cortex during the development of *aerenchyma* does not impair radial ion transport (Drew and Saker, 1986). During this development there is a drastic collapse of the cortex to provide large air channels in the root, leaving only narrow bridges of cells (Figure 5.3). *Zea mays* is a species with the facultative ability to develop aerenchyma in conditions of oxygen deficiency in the rooting medium. In spite of the dramatic decrease in the surface area of the cortex, ion transport is not impaired. At higher external ion concentrations the importance of the cortex in uptake is likely to increase as more ions reach the cortical cells.

5.3.1 *Net transport*

In previous chapters we have described vesicles, and organelles such as mitochondria and plastids, which are enclosed, defined spaces in which solute concentrations build up. As we move on to look at uptake by roots into plants we must recognise that, for most of the solutes taken up, a cell in the root is one stage of a journey; it is not a destination. There is a net, directional flux of solutes through the root, even though the instantaneous concentration of any solute in the root may remain constant. Thus we can deduce little about the transport capacity and transport characteristics of the root unless we are able to measure the rates of uptake into the root, movement to the conductive tissues of the xylem, and perhaps efflux from the root, of solutes in circumstances where the total quantity of these solutes in the root system does not change.

A simple analogy may help to explain some of the problems to be addressed. Passengers continually arrive at and depart from an airport. The number of people in the airport during the day might be more or less constant, even though individual passengers arrive and depart. It is this overall population that we measure if we take a sample of root, extract it (in dilute acid for instance) and determine its ion content. However, despite the overall constancy, there is a directional flux through the system, and this directional flux is for many purposes more important than the average number of solute molecules present at one time. There are two ways to measure the rate at which passengers arrive if all we can do is count them. If departures are slowed down or held up we witness an increase in the numbers of passengers within the terminal, because they keep arriving but do not leave. Inhibitors of transport into the xylem or stoppage of transpiration will act in this way. Alternatively, we could look at the increase in the numbers of passengers in the terminal early in the morning when it was below capacity. Measurements of the change in concentration of a nutrient on supply to a nutrient-deficient plant are analogous. In both cases we would be assuming that passengers arrived at the same rate as they would do in the middle of the day when there were no delays. If we block transport to the shoot, or use low-salt (salt-starved) roots then we can infer a rate of influx from the rate of increase in the total quantity of ions within the root. We would have another option if we made individual passengers identifiable, such as by labelling newcomers in some visible way. This is the basis of tracer studies, usually involving a radionuclide, and can enable us to measure the rate at which new solutes enter even though the total number of solute molecules in the root does not increase. We can also measure the net transport through the system by studying, at intervals, the quantity (the product of concentration in the plant and the biomass of the plant) and dividing this by the average weight of roots during the interval. A particular advantage of the use of labels is that major perturbations to the fluxes and possible consequent feedback control can be avoided. In the airport analogy it is easy to imagine that if departures were prevented action may be taken to stop the entry of passengers into the airport, thus altering the normal influx.

5.3.2 *Concentration dependence*

As the concentration of a nutrient ion in the external medium increases, so does its rate of uptake by the plant. However, as the earlier example of plant potassium concentrations (Figure 2.13) suggested, the relationship between uptake and external concentration is not linear. As presented earlier, this was an incomplete story because it concerned only concentration in the shoot, and not *quantity*; the product of concentration and biomass. However, not only the internal concentration, but the rate of uptake of a solute usually approaches a plateau as the external concentration is increased (Figure 5.4). This has the appearance of saturation kinetics and it has been common practice to take this analogy further and to interpret the saturation curve with the formalism of the Michaelis–Menten model of enzyme kinetics. This analogy has been applied for two reasons. First, it provided a convenient model to describe the data. Second, a carrier in the membrane could be thought of as an enzyme and the nutrient as its

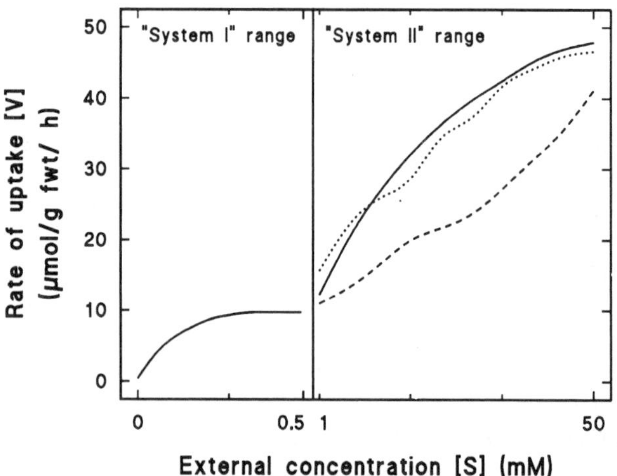

Figure 5.4 The concentration dependence of nutrient uptake in plants. The figure shows the consequences of changing the external concentration of an ion on the rate of its uptake for models based on: dual isotherms (solid line); a carrier such as envisaged in the dual isotherms model operating at low concentrations, but with uptake limited by diffusion at higher concentrations (dashed line) and the multiphasic model of Nissen (dotted line) (see text for further explanation).

substrate. The constants K_m and V_{max} have been used extensively in the literature to describe the apparent affinities and maximal velocities of ion transport processes.

5.3.3 *Dual mechanisms and multiple isotherms*

In the 1960s, descriptions began to appear of experiments in which the concentration dependence of ion uptake was examined. Excised tissue, commonly roots, was incubated in aqueous solutions of differing concentrations of salts and the rate of uptake of an ion determined. For example, the rate of uptake of potassium (often measured as ^{86}Rb) was determined from different concentrations of potassium chloride. The rates of uptake tended to saturate (to show saturation kinetics) over a particular concentration range and then to rise again to a new plateau at a higher concentration range (Figure 5.4). The uptake kinetics could be separated into a 'low' and 'high' range and were said to display a 'dual isotherm' (see, for example, Epstein, 1972). These were referred to as two distinct types of uptake mechanism, systems I and II (Figure 5.4). In the kinetic analogy, these were characterised by their affinities and maximal velocities. A considerable literature built up in subsequent years as to the nature and location of these systems, particularly as to whether they were both at the plasma membrane (parallel model) or whether system II was at the tonoplast (series model). The analysis has been taken in another direction by Nissen, who considers that uptake over an extended range of concentration is better described by a family of isotherms ('multiphasic uptake', see Figure 5.4). His analysis differs also in concluding that all the different phases take place via one, and only one, transport system (Nissen, 1991).

The kinetic analysis model has received considerable criticism. Wyn Jones (1975) presented a damaging list of arguments which centred on two areas. The first is that the confidence in individual data points obtained in the early experiments was just not good enough to allow a statistical decision to be made between rival versions of curve-fitting. The second is that there is neither good reason to believe that active transport processes have to follow Michaelis–Menten isotherms, nor that other types of process mediating ion uptake should not display saturation kinetics.

Modes of transport across the membrane, other than those mediated by carriers, may saturate for various reasons. Only so many ions can pass through a channel in unit time, for example, and there will be a point at which conductance becomes a function only of the gating of the channel. For passive diffusive processes, the driving force is the concentration *gradient*. Only initially and where the internal concentration of the solute is zero is the gradient linearly proportional to the external concentration. If the solute is not removed from, or metabolised in, the cytoplasm as fast as it diffuses across the plasma membrane, the concentration in the cytoplasm will rise, the concentration gradient across the plasma membrane will decrease, and the rate of diffusion will also decrease.

There are now grounds for seeing a mechanistic distinction between transport processes operating at different external concentrations. When external concentrations are low, the solute may not have a free energy gradient into the symplast: if it is to be accumulated this must be via a mechanism which in some way transfers energy directly from the hydrolysis of chemical bonds or indirectly from the free gradient of another solute. As the external concentration is increased, particularly for cations, a free energy gradient may exist into the symplast and all that is needed to mediate their entry is for an ion channel to open. How rigid these distinctions are is not yet clear. Some evidence associates a K-channel with the ATPase (see section 3.4.1) and Nissen (1991) views carrier and channel activity as different manifestations of the same system occurring at different concentrations.

Carrier-kinetic analysis has been one approach to the understanding of the mechanisms of solute transport into cells and has dominated the literature over a period in which the major method of experimental study, radionuclide-labelling, was peculiarly suited to the experimental approaches it called for. Unfortunately, the dividing line between a descriptive model and the inference of mechanism has too often become blurred. Three lines of contemporary investigation which we have described previously should give different insights: patch-clamping, *in vivo* molecular probes for cytoplasmic ion concentrations, and molecular studies of transport proteins.

5.3.4 *Balance between net influx to the symplast and loading of the xylem*

We have much less information on ion transport for intact plants than for excised roots. This largely reflects the ease with which experiments can be conducted using isolated roots or parts of storage tissues in comparison with those involving whole plants. Short-term uptake experiments using detached roots are usually designed specifically to isolate the unidirectional influx across the plasma membrane. In an intact plant this is only part of the story.

Let us consider the uptake of potassium by a plant growing actively in nutrient-sufficient conditions. Once the cells of the root cortex and epidermis have expanded they have no net demand for potassium, but the potassium required by the shoot (and the expanding regions of the root) passes through them. At some point, virtually all the potassium destined for the shoot must pass through the root symplast. We now know that the concentration of potassium in the cytoplasm is maintained within close limits as metabolic processes such as protein synthesis have, for their activation, precise requirements for potassium ions (Leigh and Wyn Jones, 1984). Thus potassium uptake by the root necessitates the net movement of potassium from outside to xylem with a minimal change in cytoplasmic concentration. In general, the net influx at the epidermal and cortical plasma membranes (that is the difference between influx, J_{oc}, and efflux, J_{co}, namely $J_{oc} - J_{co}$) must equal the flux from symplast to xylem within the stele (J_{cx}) if this pseudo steady state is to be preserved. The vacuoles of the root are very large in relation to the volume of the cytoplasm and can thus buffer concentration changes in the symplast by rather small changes in their own concentration and osmotic potential. In the long term though, net uptake by the root will have to match J_{cx}. We know very little about this flux.

5.3.5 *Release to the xylem*

There has been debate as to whether active transport is or is not needed to load the xylem. A single active step, the pump-and-leak hypothesis, is feasible if nutrients are at a lower electrochemical potential in the xylem than they are in the symplast. Since the ions may have been accumulated actively into the symplast, they may be at a higher electrochemical potential in the xylem than in the external medium. Measurements of radial profiles of electrochemical potential across roots made by driving microelectrodes, including ion-sensitive microelectrodes, through successive cell layers supports the idea that ions move down electrochemical gradients into the xylem (see Bowling, 1981). In no case investigated was the flux of nutrient ions to the xylem (J_{cx}) against an electrochemical potential gradient (the recordings made by Bowling and his colleagues were presumeably of conditions in the vacuoles and so strictly we do not know the electrochemical potential between symplast and xylem).

Some data are consistent with the involvement of a second regulatory step in movement from outside to xylem. Roots may be inserted into boxes that are divided up into compartments so that particular zones of the root can be labelled with radionuclides, and the exudate from the cut xylem collected. In this kind of apparatus it is possible to make measurements of uptake into the root and into the xylem at the same time, in the same root. In the short term, it has been possible to inhibit exudation of ions from the xylem independently from uptake into the cortex using inhibitors and plant growth regulators (Lüttge and Higinbotham, 1979). These observations need not be contradictory if we envisage the movement from the symplast of the xylem parenchyma into the xylem as occurring through channels. Loading the xylem would be capable of separate regulation from uptake, but would be down a free energy gradient. There has long been a view that ions leaked out of the symplast within the stele into the xylem vessels; a model proposed by Crafts and Broyer in 1938. The tissue of the stele was considered deficient in oxygen, as a consequence of which the cells of the stele were inherently leaky. However, Bowling (1973) did not observe a serious reduction in the partial pressure of oxygen within the stele when he measured the radial

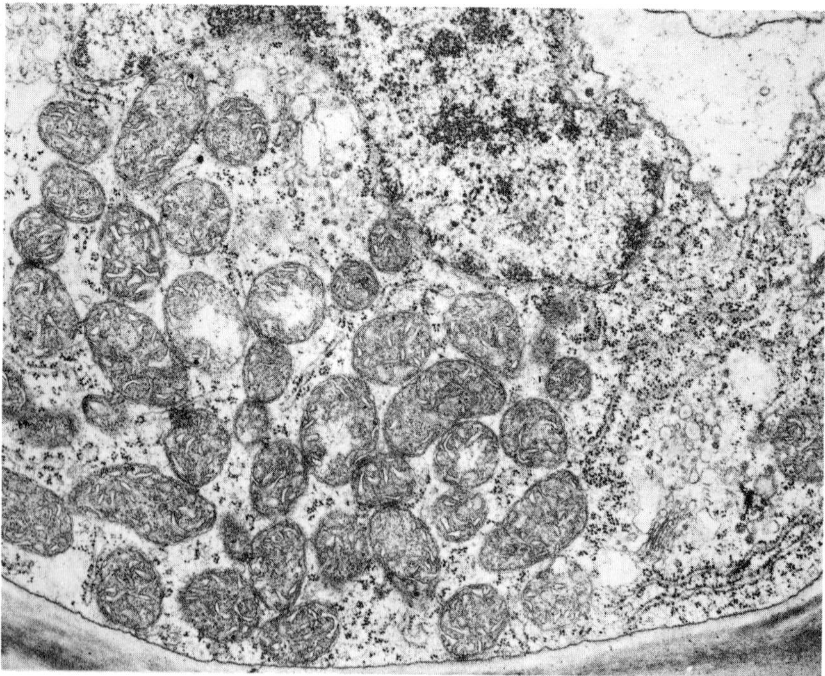

Figure 5.5 An electron micrograph of a cell from the xylem parenchyma of a mature maize root. Note the polysomes, rough endoplasmic reticulum and preponderance of mitochondria. Magnification × 24 300.

profile with an oxygen sensitive microelectrode. The identification of ion channels in plant cell membranes now provides a mechanism to explain passive movement from symplast to xylem without the need to propose that the membranes leak because they are impaired metabolically. Cytoplasmic homeostasis with respect to potassium (see Leigh and Wyn Jones, 1984) would be problematic if the symplast did uncontrollably leak its contents to the xylem. The appearance of xylem parenchyma cells in electron micrographs, densely cytoplasmic with numerous organelles, is at least circumstantial evidence that they retain metabolic integrity (Figure 5.5). Xylem parenchyma cells also have been ascribed active roles, particularly in removal of toxic ions such as sodium, from the xylem. There are also data which are suggestive of electrogenic proton-pumping into the stele (see Clarkson, 1988). However, solute release to the stele remains an area which is poorly understood. The lack of knowledge is mostly due to methodological limitations. Not only do we need to know ion activities within the cytoplasm (these measurements are discussed in section 1.2) but we need to know this in very inaccessible places deep within the root where visualisation has so far been impossible.

5.4 Movement across the root

There are three ways in which water and solutes can move across the root (excepting the exodermis and endodermis where there are severe restrictions). These are the apoplastic, symplastic and transmembrane pathways (Figure 5.6). The arguments presented above suggest the conclusion that most ions are absorbed across the epidermal plasma membrane. From then on they could remain in the symplast or move from cell to cell across the intervening membranes. All three pathways may be involved in water movement to some extent (see Taiz and Zeiger, 1991). A mathematical analysis of the resistances of different pathways of movement across the root was made by Tyree (1970). The driving force needed for diffusion through the plasmodesmata is remarkably small and the symplastic pathway has a much greater conductance than it might be credited with by looking at electron micrographs. Tyree (1970) calculated that a concentration difference of just 0.1 $mol\,m^{-3}$ was sufficient to drive diffusion across a plasmodesma. This would need to be increased to take account of the proportion of the overall pore cross-sectional area which more recent electron microscopical studies have shown is partially blocked (see Figure 4.12), but we are still dealing with very small diffusional driving forces and

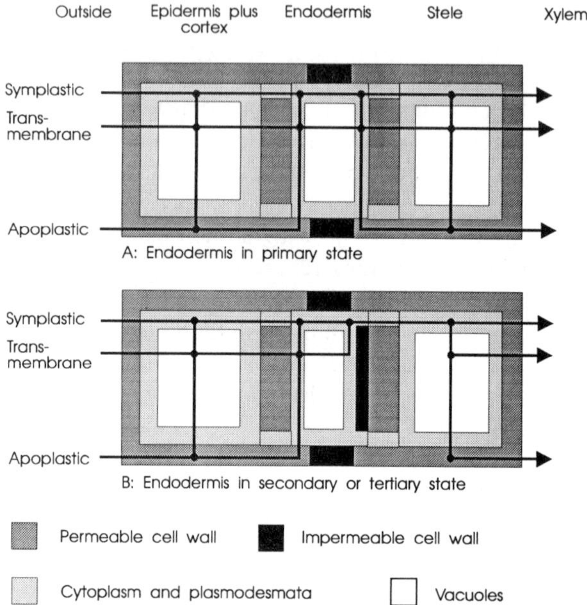

Figure 5.6 A schematic diagram of the pathways available for the passage of solutes across a root from the external solution to the xylem. The three routes are apoplastic (in the cell walls, interrupted at the endodermis and perhaps exodermis), symplastic (via the plasmodesmata) and transmembrane (a pathway that takes the solute across successive cell walls and plasma membranes).

hence a very conductive pathway. The uptake of potassium and phosphate is unhindered, even by the presence of a tertiary endodermis where the only pathway was plasmodesmatal (Lüttge and Higinbotham, 1979), further indicating that the symplastic pathway has the capacity to account for radial transport of these ions. Ion movement would be by diffusion across the plasmodesmata and by diffusion which could be aided by cyclosis across the cells. The inhibitor of cytoplasmic streaming, cytochalasin B, did not reduce potassium (^{86}Rb) transport across the root (Glass and Perley, 1979). This single investigation would suggest that diffusion through the cytoplasm is not rate-limiting and that despite the frequency of plasmodesmata (see section 4.4.2) diffusion from cell to cell is the rate limiting step. The observations of the effects of the development of aerenchyma on ion transport by Drew and Saker mentioned earlier also emphasise that, in normal conditions, the symplast has enormous reserve transport capacity, because severe loss of cellular connections across the root does not reduce radial transport. Interactions with the flow of water through the system are also possible. At the endodermis, all radial transport, whether of water or solutes, must cross the endodermal cell membranes to pass the Casparian band and, in later states of development, must cross the endodermis via the plasmodesmata. Consequently, the symplast must be capable at some point of sustaining the total water and solute transport of the root.

The hydraulic conductance of single cells has been measured with the miniature pressure-probe (Steudle and Jeschke, 1983). This, together with kinetic studies, now suggests that the apoplast is not a major pathway for radial water movement across the root (Boyer, 1985) and therefore that most of the water moves symplastically or from cell to cell across the membranes as far as the endodermis. The apoplast may increase in significance for the movement of water at high transpiration rates and for ions at high external concentrations. Apoplastic movement can also increase in importance, especially in relative importance, in certain stress conditions such as anoxia caused by waterlogging. This decreases the hydraulic conductance of both the plasma membrane and the plasmodesmata.

5.4.1 *Effect of the state of development of the root*

We have seen that ions are transported across the root irrespective of the state of development of the endodermis. However, endodermal development poses serious constraints for the transport of some ions, calcium in particular. The calcium concentration in the cytoplasm is generally below 400 $\mu mol\ m^{-3}$ and changes in its concentration have major regulatory roles. The calcium requirement of a plant could not possibly be met by the quantity of calcium that could be transported symplastically inwards at a concentration that would not interfere with these regulatory functions. Therefore, radial calcium transport cannot be symplastic. This accords with the correlative data between the uptake of calcium and the stage of development of the endodermis (see Figure 5.2): calcium can only move into the root in appreciable quantities before the endodermis blocks effectively the apoplastic pathway. Since calcium is the second most abundant cation in plant cells (see section 3.4.3) and does not cross the root in the symplast, this implies that the apoplast has considerable capacity for radial transport, even though the consensus is that most ions move symplastically.

5.4.2 *Limiting factors in ion absorption*

Agricultural improvement may be seen to require increases in the efficiency of nutrient acquisition by crop plants but the pathway by which ions are accumulated by plants is a complex one, so it is not easy to guess what is limiting and consequently where improvement is needed. Although the characteristics of ion uptake and the conductance of various pathways have had a prominent place in the literature, neither the relative conductances of the symplast and apoplast, nor the transport capacity at high external concentrations, is necessarily limiting. The root has excess, unused capacity in all these areas. Ions may mostly move in the symplast because supply in the soil is usually limiting and such nutrients as are available are scavenged at the root surface. We also have a very uneven picture of plant mineral nutrition as Clarkson (1985) emphasises strongly. Most of the data we have on the characteristics of ion uptake by roots derive from seedlings (to which the roots may no longer be attached) and are clean, straight structures, usually devoid of root hairs, grown in a nutrient solution. This solution is usually stirred, aerated and replenished at intervals to maintain a high and fairly constant concentration. In order to maintain a perspective it is important to recognise that additional limiting factors are involved when plants are growing in their natural environment.

5.5 Addendum

5.5.1 *Donnan systems*

The cell wall represents a solid phase with fixed negative charges: similar systems are to be found in the soil and in membranes. They are known as Donnan systems and the concentration of ions within water-filled pores in the solid phase differs from the concentration in the bulk solution with which the solid phase is in contact and in equilibrium. Calculation of the concentrations may be complex, but two examples will serve to illustrate the difference that can occur between Donnan phase and solution.

In a Donnan system, such as the cell wall, where an ion is in equilibrium between the solid phase and the solution,

$$\mu_{(\text{solution})} = \mu_{(\text{cell wall})}$$

or

$$\mu_{\text{s}} = \mu_{\text{cw}}$$

hence,

$$\mu_j^* + RT \ln a_{j\text{s}} + z_j F E_{\text{s}} = \mu_j^* + RT \ln a_{j\text{cw}} + z_j F E_{\text{cw}}$$

The μ_j^* terms will cancel and an equation relating concentrations or activities and electrical potentials can be developed, from which the following examples were calculated by Briggs *et al.* (1961).

The fixed negative charge in the cell walls is 100 mol m^{-3}. The external concentration of KCl is 10 mol m^{-3} and in the second case, there is also 1 mol m^{-3} $CaCl_2$ present:

A^-	100	mol m^{-3}
KCl_0	10	mol m^{-3}
Cl^-_{cw}	0.99	mol m^{-3}
K^+_{cw}	100.99	mol m^{-3}

In the presence of 1 mol m^{-3} Ca^{2+}

A^-	100	mol m^{-3}
KCl_0	10	mol m^{-3}
Cl^-_{cw}	1.84	mol m^{-3}
K^+_{cw}	62.5	mol m^{-3}
Ca^{2+}_{cw}	39	mol m^{-3}

CHAPTER SIX

TRANSPORT FROM ROOT TO SHOOT

The movement of solutes into and across the root described in the last chapter suggested little interaction between the flows of solute and solvent. From the root to the leaves, however, the flow of water and solutes is closely coupled. The xylem has evolved to transport the large volumes of water needed to replace transpirational losses and presents a system of large capacity and speed which provides a 'free ride' for anything that will dissolve in water. The xylem, however, is a one-way pathway from the roots to the leaves, and its capacity and exact destination are determined by how fast particular parts of the shoot are losing water to the atmosphere, not by the requirements of the shoot for solutes. As a system for solute transport this does have shortcomings.

6.1 Structure and function of the xylem

6.1.1 *Structure*

The xylem is the single most striking feature of the anatomy of a mature root. In cross-section cavernous holes are seen in the central region (the stele) of the roots of angiosperms (Figure 5.1). In longitudinal section these holes can be seen to be the lumina of long, straight pipes with a degree of strengthening which suggests they may have evolved to withstand substantial deformational forces (Figure 6.1). The large diameter and armoured construction to the xylem reflects the harsh reality of life out of water.

The xylem of most angiosperms is a mixture of *xylem vessels* and *tracheids*. Gymnosperms generally lack vessels. Vessels are typically of larger diameter than tracheids and are formed from a series of cells connected end-to-end via *perforation plates* (in which the wall has holes), as well as to adjacent vessels and tracheids through pits (regions of permeable, primary cell wall in the impermeable, lignified wall). Tracheids are single,

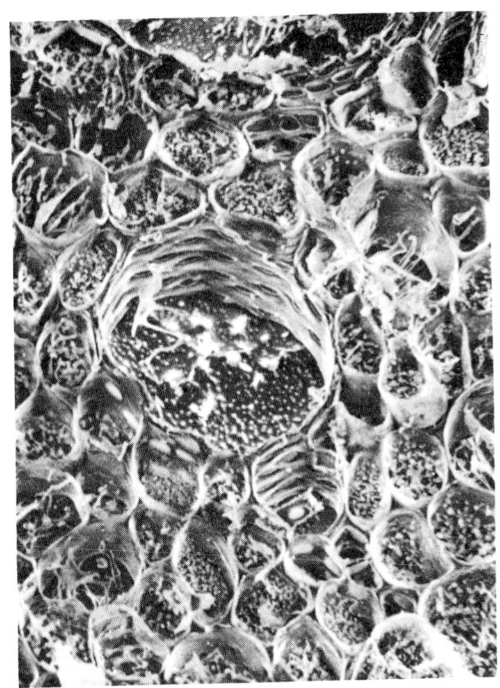

Figure 6.1 Scanning electron micrograph of the freeze-dried surface of a root of *Zea mays* showing a large and a small vessel element with characteristic strong, lignified walls and a large number of prominent pits (areas of unthickened primary wall which remain permeable to water). Magnification, × 750.

elongated, lignified cells interconnected by pits in all directions, but not formed into longitudinal pipes.

6.1.2 *Water flow in the xylem*

The leaves of most plants have a large surface area and the epidermises are perforated by stomata. The number and aperture of stomata has to be sufficient to allow carbon dioxide to diffuse into the leaf to satisfy the requirements of photosynthesis. The concentration gradient down which this takes place is about two- to four-fold; the larger values occurring in C_4 species which have lower intercellular carbon dioxide concentrations (C_i) than C_3 plants. The *absolute* concentration of carbon dioxide in air is, however, low, so the driving force for gaseous diffusion inwards, the difference in partial pressure of carbon dioxide in the air outside (C_a) and inside (C_i) the leaf, is small and typically 10 Pa in a C_3 plant. Carbon dioxide and water vapour follow the same diffusional pathway, albeit in opposite directions, the only difference being that the smaller water molecules diffuse about 1.6 times as fast as the carbon dioxide molecule. Beneath the stomata the cell walls are saturated (or almost so) with water. Since the surface area of these cell walls is very great in comparison with the volume of the intercellular spaces it is assumed that the air in the intercellular spaces is also practically saturated with water: At typical leaf temperatures, the partial pressure of water vapour in the intercellular spaces (e_i) would therefore be thousands of Pascals. In contrast, in most parts of the world in daylight the partial vapour pressure of water in the atmospheric air (e_a) is very far from saturation. So, although the concentration gradient of water vapour is only a few-fold (except in very dry air where it is larger), the *absolute* concentrations and partial pressure differences for water are very much greater than for carbon dioxide. Taking a specific example (Taiz and Zeiger, 1991), for a leaf temperature of 30°C and an average atmospheric relative humidity of 50% (atmospheric air containing half the amount of water it could carry in the vapour phase at that temperature), the vapour pressure difference for water is 2.1 kPa. Comparing this with the vapour pressure difference for carbon dioxide of just 10 Pa, and allowing for the relative diffusion coefficients, the plant is expected to lose over 300 molecules of water for each carbon dioxide molecule that diffuses in ($2100/10 \times 1.6 = 336$). This flux of water may be three times the volume of water in the leaf each hour (Boyer, 1985).

The loss of water from the leaf that results from the need to have a free diffusive path for carbon dioxide entry to the leaf, necessitates the continual replacement of leaf water otherwise the leaves would rapidly desiccate and die. This replacement is supplied through the xylem. Dead, hollow cells, and in particular the wide-diameter xylem vessels, increase dramatically the axial conductance to water flow. In order to evaluate the importance of pore diameters to flow through the xylem it is necessary to understand the association between flow, diameters, pressure and viscosity. The relationship was first described in the mid-1800s, by Hagan in 1839 and by Poiseuille in 1840 and is often known as Poiseuille's law:

$$J_v = \frac{-r^2 \Delta P}{8\eta l}$$

The equation relates the volume flowing per unit time per unit area, J_v, in a cylinder of radius r and with a difference in pressure, Δp, over a distance, l, driving the flow; η is the viscosity of the solution that is flowing (see Nobel, 1983). Vessels are not smooth pipes, so flow across the end-plates and turbulence could lead to some over estimation of flow per unit pressure difference, but there is good agreement between the estimate based on the Poiseuille law and experimental observation for a vessel of 20 μm radius. The value for the flow is -0.02 MPa m^{-1} to overcome resistance to flow in the tube, plus an additional -0.01 MPa m^{-1} to overcome gravity (Nobel, 1983; section 2.8.2). The key feature of such axial flow is that flow increases (or the pressure needed to drive unit flow decreases) in proportion to r^4. The flow per unit cross-sectional area increases as r^2 and the cross-sectional area over which flow is taking place also increases as r^2. Thus, for the same driving force (pressure difference) doubling the radius of a xylem vessel will increase the quantity of water flowing in unit time by a factor of 2^4 or 32-fold. Peak velocities for xylem movement in trees are typically several millimetres per second and can be 13 mm s^{-1} (45 m h^{-1}) in trees with large vessels of radius 200 μm (Taiz and Zeiger, 1991). At the extreme, wheat plants can grow with only a single seminal root axis, in which case they have essentially a single xylem vessel to conduct all the water. The mean speed of water in the xylem is 0.8 m s^{-1} (Passioura, 1988). Plants clearly need water pipes, and they need large ones, but why are they so heavily strengthened? To answer this question we must consider why water rises in the xylem. Water moves upwards against gravity, turbulent and frictional resistances to flow and some force must push it from the roots to leaves or pull it to the leaves from the roots.

6.1.3 *Driving force for water movement in the xylem*

Positive pressures (root pressure) can be generated in the xylem and demonstrated as the eventual exudation of solution from a detopped plant. Under conditions of good water supply, but otherwise poor conditions for transpiration (when there is only a very small gradient between e_i and e_a), many plants exude fluid, especially from their young leaves, which can be mistaken for dew. This exudation, which is known as *guttation*, occurs from specialised collections of thin-walled cells, known as *epithem*, which are well supplied with xylem elements and underlie the epidermis. The epidermal cells are generally modified to form a pore similar to a stoma and the whole structure is known as a *hydathode*. High pressure in the xylem (root pressure) forces water through the pore. It is thought to be important as a substitute for transpiration in the movement of solutes from roots to shoots under conditions where transpiration is insufficient to fulfil that need.

Guttation does not appear to involve metabolic activity of the epithemial cells.

Root pressure exudation occurs because the concentration of solutes in the xylem is greater than it is in the external medium and therefore there is a gradient in water potential from the outside to the xylem down which water can flow passively. This is not, however, how water usually gets to the tops of plants. If it was, then the water in the xylem would be under pressure and if the shoot was cut both cut ends would exude xylem sap. On the contrary, if the stem of a plant outdoors in daylight is cut, the xylem of both surfaces dries instantly as water is sucked away from the cut surfaces. The water can be forced back out through the xylem by putting one end under pressure. This phenomenon is used in a method of determining plant water status with a piece of apparatus called a *pressure bomb* (Figure 6.2). The pressure bomb (or Scholander bomb, after its inventor) consists of a strong metal chamber into which a plant part (stem, petiole or leaf) can be sealed, with a gas-tight seal. If a shoot is placed in the chamber with the cut end protruding through

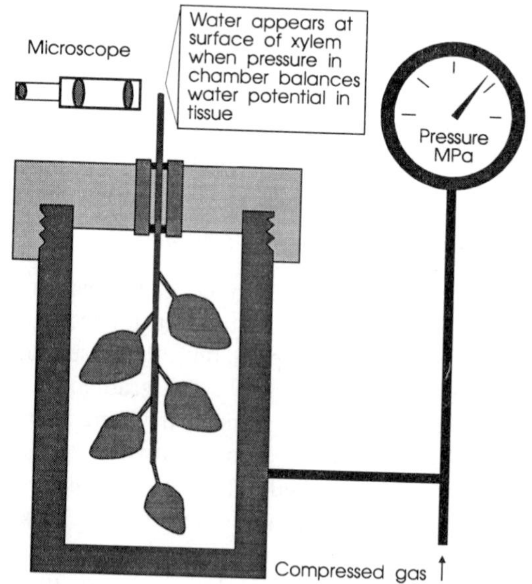

Figure 6.2 Schematic representation of the pressure bomb, often called a 'Scholander' bomb. Tissue is sealed in the container with the petiole protruding and the pressure in the chamber increased until microscopical observation shows that the cut surface of the xylem has become wetted. This, the balance pressure, is equal and opposite to the tissue water potential.

the seal into the atmosphere, pumping the chamber of the bomb full of compressed gas balances the (negative) water potential of the shoot (the water potential of the xylem is assumed to be close to the water potential of the whole organ) and water reappears at the xylem surface. The pressure needed to achieve this may typically be 1 MPa and hence the water potential of the xylem is commonly −1 MPa. If the xylem fluid is analysed it is found that its *osmotic* potential is typically −0.1 MPa and so accounts for little of the negative water potential of the xylem of a transpiring plant. The cut xylem surface dries because water in the xylem is under *tension*; not pressure. The terminology here can be confusing. In the equation for water potential the pressure component (P) will have a negative sign because it is less than the standard pressure (atmospheric) used in the definition of water potential (section 2.8.2) and the xylem is often said to be under *negative hydrostatic pressure*. We will use the term tension rather the 'negative pressure': in an absolute sense there cannot be negative pressure.

The tension in the xylem is a demonstration that the osmotic forces that lead to root pressure exudation do not transport water into the xylem rapidly enough to replace transpirational losses; only under conditions of very low transpirational demand is a pressure in the xylem manifest. Water moves up the xylem by gradients in tension.

The evaporative loss from the leaf surface is converted into a tension in the xylem because of the intramolecular forces that lead to *surface tension*. The extensive hydrogen bonding between water molecules attracts them together (*cohesion*) and to other surfaces such as the glass of a beaker or the wall of a cell (*adhesion*). A consequence of these interactions is that the free energy of a system is lowest with a minimum surface area. A free energy change associated with a change in surface area can thus provide the driving force for movement. Surface tension becomes more important the smaller the object gets. This is because volume changes as a function of r^3 and surface area as a function of r^2. Thus, the smaller an object becomes, the greater is the ratio of its surface area to its volume. The surface tension of a bathtub full of water is small in relation to gravitational effects (it has a surface to volume ratio of little more than 10 $m^2\ m^{-3}$), and the surface is flat except for a small region of *meniscus* where the water surface curves as it adheres to the edge of the bath. For a volume of water equivalent to a small plant cell (radius 10 μm) the surface to volume ratio is 300 000 $m^2\ m^{-3}$ and such a volume of *free* water would be pulled by surface tension into a spherical droplet. As far as the transpiration stream is concerned, it is the meniscus formed by the adhesion of water to cell walls that is important. The pressure (P) is inversely proportional to the radius of curvature (r) of a

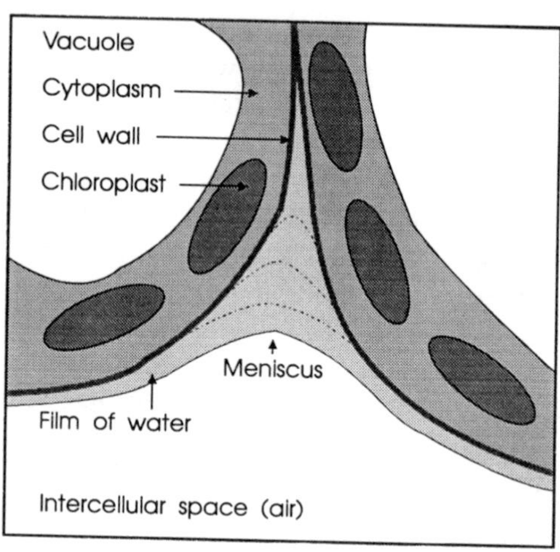

Figure 6.3 Diagrammatic representation of the water film at the junction of two mesophyll cells. A film of water covers the leaf surface and forms a meniscus at the cell junction. As the leaf dries, the water recedes into the interstices, the radius of curvature of the meniscus becomes smaller (shown by the dotted lines), and the tension in the water increases.

meniscus according to the equation

$$P = -2T/r$$

where T is the surface tension of water (0.072 Pa m at 25°C) and r is the radius. As the radius of curvature decreases, which it does when the leaf tends to dry and water is withdrawn into the crevices between cell walls and eventually into the interstices of the cell walls themselves (Figure 6.3), so the tension increases. When the radius of curvature is as little as a few micrometres, the magnitude of the resulting tension becomes physiologically significant (at 1 μm, P is -0.15 MPa and at 10 nm, P is -15 MPa).

This tension is transmitted from the cell wall surface to the xylem so that the solution in the xylem is under tension. The hydrogen bonds that give rise to the cohesion of water molecules also provide the tensile strength of water, so that even if it is under a tension of several million pascals, a water column does not break. The empirical value for the tensile strength of pure water in capillaries is 30 MPa (Nobel, 1983), though this is reduced drastically by impurities, particularly by gases. The latter are the cause of *cavitation* or breakage of xylem water columns. It is relatively simple to calculate from the Poiseuille equation that a pressure gradient of 3 MPa is needed to carry water to the top of a 100 m tall redwood tree (*Sequoia sempervirens*). For a velocity of 1 mm s^{-1}, a viscosity of 1.0 mPa s and a radius of 20 μm, the pressure gradient necessary to drive the flow through the xylem vessels is 0.02 MPa m^{-1}. Since it requires 0.01 MPa m^{-1} to overcome gravity (section 2.8.2), the total pressure required is 0.03 MPa m^{-1}. This will be almost entirely tension.

Hence the answer to our second structural observation on the xylem; it is heavily strengthened to carry water under tension without collapsing.

6.1.4 *Water movement in the leaf*

On reaching the leaf, water, and the solutes that it contains, is first distributed in xylem vessels. The volume flow in vessels increases dramatically with radius and so the large vessels carry practically all the water. The xylem of the leaf can be viewed as a system of large supply veins into which a network of smaller distribution veins are tapped; large veins convey water rapidly across the leaf to sites of maximum evaporation, which may be at the leaf margins where conductances through the boundary layer are greatest (Canny, 1990).

Most water evaporates from inside the leaf and passes out through the stomata. Exactly where inside the leaf most water evaporates from is not fully resolved. Some models put the site of maximum evaporation close to the stomata (see Boyer, 1985). Other data put the evaporation site much deeper within the leaf. Farquhar and Raschke (1978) used leaves which had stomata on both surfaces and supplied helium to one side of the leaf and air to the other. They compared the diffusion of helium *through* the leaf with that of water *out* of it (water would need to cross at most half the leaf thickness to reach the epidermis). The diffusive resistance to water loss was almost half the resistance to helium diffusion through the leaf. This leads to the conclusion (Boyer, 1985) that evaporating sites are deep within the leaf close to the vascular system.

The question then remaining is how water moves from the xylem to the evaporating surface. Historically, opinion has swayed between apoplastic and symplastic pathways. The view at present is that it is not apoplastic (Boyer, 1985; Canny, 1990) but there does not yet appear to be any experimental evidence on which to apportion symplastic and transcellular movement across the intervening cells. The hydraulic conductance of leaf mesophyll cells appears to be similar to most other higher plant cells (10^{-9} m s^{-1} MPa^{-1}) and the conductance of the apoplastic pathway appears to be too small to account for much water movement (Boyer, 1985). The water potential gradient needed to support the transpirational flux of water across a single mesophyll cell in a rapidly transpiring sunflower plant was calculated by Boyer to be 0.6 MPa. The water flux could not pass through many cells in series for a realistic gradient in water potential, which also implies that evaporation sites are close to the vascular tissue.

As a consequence, most of the water is evaporating from a fraction of the cells of the leaf and the flux of water through cells close to the veins may be such that the water content of these cells has a turnover time of 2 minutes (Boyer, 1985). Canny (1990) summarises microscopical evidence obtained with membrane-impermeant fluorochromes. These are separated from the transpiration stream where the water passes from apoplast to symplast and therefore a localised accumulation of these tracers indicates sites at which water enters the symplast. These tracer experiments also suggest that the transpiration stream enters the symplast at a point very close to the tracheary elements; furthermore entry occurs over small local areas of membrane where the flux of water will be even greater than the above calculations suggest. Precise localisation of the sites at which water enters the symplast can avoid problematic interaction with sucrose movement towards the phloem (chapter 7). In wheat leaf veins, it appears that water from the xylem enters the symplast of the upper parenchyma cells, while sucrose enters from the lower side (Canny, 1990).

In all cases, the transport of water through the plant is driven by the evaporation of water to the atmosphere. The transpiration stream requires no direct input of resources by the plant. Indirect costs include synthesis and maintenance of the conductive pathway.

6.1.5 *Consequences for solute transport*

Liquid water moves to the leaves in the xylem, a part of the apoplast specialised for conductance. In the leaves in daylight most of this water passes into the vapour phase. Although the water ultimately evaporates from the apoplast, it has probably passed *through* few living cells between the xylem and the site of evaporation. Since the water in the xylem typically contains dissolved solutes with an osmotic potential of -0.1 MPa (equivalent to about 40 mol m^{-3} sucrose) and most of the water is lost as pure water, what happens to the dissolved solutes? They must be separated from the transpiration stream at some point between the xylem vessel entering the leaf and the evaporation site.

From the point of view of leaf cells, the xylem sap is the 'external' solution in which the cells are bathed, but we have little knowledge of the composition of this solution as it is not particularly easy to obtain. The

Table 6.1 The concentration of some inorganic solutes in the xylem sap of sunflower (*Helianthus*). Data are the range of concentrations obtained in the xylem for soil water contents from 0.15 to 0.20 g water per g dry weight of soil (well-watered conditions) (Schurr and Gollan, 1990)

Solute	Concentration (μM)
Nitrate	5000–15000
Phosphate	200–700
Calcium	300–1200
Potassium	2000–8000
Magnesium	200–800
Sodium	20–100
pH	5.8–6.6

analyses available show the solution in *Helianthus* to be dominated by nitrate and potassium ions and to have a pH on the acid side of neutral (Table 6.1). In a flowering maize plant the potassium is balanced mainly by organic acids and also contains 5 mol m^{-3} sugars (Canny, 1990). Although measurements may be obtained of the solution in xylem exudate, the average contents of the xylem vessels are as remote from the cell walls immediately surrounding the leaf cells as the bulk soil solution is from the cortical cells of the root (cf. section 5.1.1). Direct measurement of the ion concentrations in the apoplast of leaves is technically extremely difficult (much more difficult than obtaining xylem exudate) and results available are too few for generalisation. From an experimental point of view it is possible to measure uptake into leaf discs and slices from solution. However, while diffusion from the bulk solution is a fair model of how ions get to the absorption sites of a root, it is clearly not a fair model of how they reach them in a leaf: solutes do not diffuse up the fine ramifications of the minor veins, they travel in a mass flow of solution. For these reasons we have very little specific information about the movement of ions from the xylem into leaf cells.

The deductions about the pathways of water movement through the leaf have one further important implication. The sites at which solute and solvent may become separated will be localised; for example where water enters the symplast rapidly and where evaporation is maximal. Subsequent distribution of the solutes is likely to take place via the symplast. The studies mentioned in the previous section, with membrane-impermeant fluorochromes, provide some evidence of these discontinuities. The localised accumulations of these tracers at sites of solvent–solute separation (Canny, 1990) suggest that similar effects might indeed be seen with the xylem's normal solutes. Although (unlike the impermeant fluorochromes) these solutes may cross to the symplast by any of the transport methods described earlier (carriers and channels) these transport systems may not have the localised capacity to match the localised movement of water. We might anticipate that the distribution of solutes in the leaf apoplast, and consequently the sites of uptake of these solutes, may be patchy, making experimental modelling of the leaf even more difficult.

The same microscopical studies suggest two other features of solute movement within the leaf. The first is that solutes can diffuse away from sites of localised accumulation, even against the direction of mass flow, and that the direction of the diffusional pathways may be strongly influenced by the fine structure of the cell walls along which diffusion occurs. The second is that the cells of some regions of the leaf appear to function as 'scavenging' sites to remove solutes from the apoplast (Canny, 1990).

Solute uptake in leaves represents a major gap in our knowledge of solute transport in plants. To investigate the fates of the usual solutes of the xylem will require methods with the resolution of X-ray microanalysis (because we need to know that is happening in localised regions of cell wall) but which are rapid enough to use in conducting experiments and which can handle solutes which are organic and/or composed of low atomic number elements. This is likely to remain a frontier area for some time to come. The only general statement to be made is that, as was the case with the root, it appears that the uptake and transport capacity of leaf cells is large in relation to the rate at which solutes usually arrive in the xylem, otherwise extracellular accumulation of solutes would rapidly become a problem throughout the leaf. The only circumstance in which this has been shown to occur is in saline conditions (see section 6.2.5).

There is one general problem with the xylem as a transport system. Delivery is determined largely by the size of the transpirational sink; not by the specific nutrient requirements of a particular part of the shoot. Two factors necessarily modify this: accumulation of solutes from the xylem along the transpirational pathway and movement of solutes from xylem to phloem. From the point of view of water movement we have regarded the xylem as an inert pipe but it is bordered by living cells and solutes are removed from the xylem along its course by absorption which may be preceded by adsorption. A specific example, of selective removal of sodium from the xylem stream, is described below (section 6.2.5). The transpiration stream carries nutrients from root to shoot, and carries them effectively and economically, but they need to be redistributed. The transpiration stream will carry too many nutrients to expanded leaves which no longer need them and not enough to expanding and non-transpiring organs which have the greatest requirements. The movement of solutes between xylem and phloem is dealt with in more detail in the next chapter (section 7.6).

6.2 Salinity as a model system

On many occasions so far we have encountered questions which are difficult to answer. For example, we have seen that alternative pathways for the movement of solutes (involving apoplast and symplast) may each have so much capacity in reserve that either could be responsible for all of the transport observed (see section 5.4). How do we tell which pathway is usually important? The standard experimental approach is to devise treatments or conditions which disturb the system enough to allow us to discriminate between different hypotheses describing how the system might function. However, it may need a big disturbance, or an unusual condition, to provoke a change that will saturate or block a particular transport pathway. If these perturbations are not within a normal physiological range then the outcome is at best difficult to interpret because we are trying to answer questions about the normal by extrapolation from the abnormal (i.e. conditions that never exist in nature). The study of plants that have evolved to occupy extremes in the ecological range may provide some answers.

6.2.1 *Halophytes*

Halophytes are plants whose natural habitats, coastal marshes and inland salt lakes and deserts, are saline. This means that their soils have a high concentration of soluble salts; high enough to kill the majority of plant species (so-called *glycophytes* or plants of fresh or sweet water; most plants are glycophytes). The most important salt is normally sodium chloride, the salt that predominates in seawaters. All halophytes survive in high salt concentrations (most seawaters contain about 480 mol m^{-3} Na^+ and 560 mol m^{-3} Cl^-); some simply survive, while others, mainly from families in the Dicotyledoneae, often make their optimal growth in the presence of about 200 mol m^{-3} sodium chloride. These dicotyledonous halophytes tend to have succulent leaves, containing high concentrations of sodium and chloride ions (about 500 mol m^{-3}). Monocotyledonous halophytes also contain high concentrations of ions but, with one or two exceptions, are not succulent, having lower water contents and sodium to potassium ratios than are found in the dicotyledonous species (Flowers *et al.*, 1986). Most of the ions are stored in the leaves, but all must pass through the roots. In some species, the concentration of ions is, in part, regulated by the operation of glands that remove ions from the leaves (see section 6.2.8). In others, the ion concentration within the leaves is regulated by a tight coupling between ion transport and growth. The Chenopodiaceous species *Suaeda maritima* is a model for such halophytes and has many advantages as an experimental material.

Suaeda maritima grows naturally in flooded conditions, hence growing plants in culture solution is reasonable. The Chenopodiaceae do not have mycorrhizal associations, which helps further reduce the artificiality of hydroponic culture. The plants grow, and have evolved to grow, naturally at external salt concentrations often in excess of 500 mol m^{-3}, enough to saturate both the uptake and radial transport capacity of the root. Experimental observations and manipulations are made very easily because the high concentrations in all compartments of the cells are readily measured. Growth 'at the limit' is an ecologically and physiologically relevant condition and not an arbitrary experimental treatment.

6.2.2 *Site of ion uptake in the root*

The net transport of ions can be determined from sequential harvests. In these experiments plants are collected at intervals of one or two weeks and the content of given ions in the roots and shoots measured. From this data it is easy to calculate the change in the quantity of those ions in the shoot per unit time and per unit of root weight. The maximal net transport of NaCl by *S. maritima* is about 10 mmol per gram dry weight of roots per day, which is at least ten times the net potassium transport by a well-fertilised crop plant (see Yeo and Flowers, 1986). If we divide this net transport by the surface area of membrane across which it may pass to enter the symplast then we arrive at a minimal value for the net flux into the plant. The surface area values may be obtained by quantitative analysis (stereological analysis) of electron micrographs (see Steer, 1981). The flux is equivalent to 3000 nmol m^{-2} s^{-1} at the root surface, 1000 nmol m^{-2} s^{-1} over the entire epidermal cell surface and about 250 nmol m^{-2} s^{-1} over the whole epidermal plus cortical plasma membrane surface. For comparison, the active chloride flux in the stomatal guard cells of 200 nmol m^{-2} s^{-1}

accounted for the entire ATP resources of these specialised cells (see section 3.5). It is difficult to account for active chloride uptake in *S. maritima* unless chloride has access to the cortex.

How can we tell by what mechanism ions are usually accumulated and by which pathway they cross the roots? Generally, we cannot. For the quantities transported by a crop plant, the movement of potassium is apparently unaffected by stopping cytoplasmic streaming and by loss of most of the cortex during the facultative development of aerenchyma (see sections 5.3 and 5.4). It is not possible to discriminate by experiment between the apoplastic and symplastic pathways. The quantities of ions moving in an halophyte, however, make it straightforward to calculate that there is no significant flow of water in the apoplast across the root. The solution flowing towards the root to replace the transpiration stream is not just water, or even dilute nutrients, but a saline solution approaching or equalling seawater in concentration. The molar ratio of net absorption of water and of sodium by *S. maritima* is about 1000; if there were a mass flow of solution to the endodermis the salt which was left behind by the flux of water would crystallise in the root within the hour (Flowers and Yeo, 1988). We must conclude that the flow across the root is largely symplastic.

The driving force for the symplastic movement of ions, assuming they are moved across cells by cyclosis, need not exceed 10 mol m^{-3} per cell, or 40 mol m^{-3} across the narrow cortex of *S. maritima* (*S. maritima* has only three layers of cells in its cortex), at an average symplastic concentration around 100 mol m^{-3} (Yeo and Flowers, 1986). Ion uptake by *S. maritima* is probably limited by the energy available to accumulate chloride and by the capacity of the symplast to carry NaCl across the root at a sub-lethal concentration. Above concentrations of about 150 mol m^{-3} in the cytoplasm, sodium and chloride are toxic as they disrupt protein synthesis and interfere with the activity of enzymes such as dehydrogenases. This halophyte should provide good experimental material for an investigation of the role of cytoplasmic streaming, but above all offers a system for the investigation of regulation of ion transport. The net flux is large and the pseudo steady state concentration in the symplast close to critical.

What happens when unloading at the xylem (J_{cx}) is perturbed? Is influx stopped? Is efflux increased? What is sensed, turgor, membrane potential or ion activity? We do not know, but questions might be answered here that could not be answered using a plant growing in 1 mol m^{-3} potassium. We can, for example, deduce that the uptake of sodium and chloride ions into the cytoplasm must be capable of regulation. If J_{cx} were to be suddenly inhibited by a reduction in transpiration (a natural occurrence for a halophyte on submergence by an incoming tide), uptake into the symplast must be stopped or the cytoplasmic concentrations would rapidly exceed the lethal limit. This comes about because the symplastic concentration is determined by the difference between J_{cx} and ($J_{oc} - J_{co}$; see section 5.3.4); if J_{cx} is suddenly reduced, net influx to the symplast must also be reduced (Yeo and Flowers, 1986).

6.2.3 *Ion release to the xylem*

If the entire plasma membrane surface area of the pericycle and parenchymatous cells within the stele were used to transfer ions from the symplast to the stelar apoplast, the flux of NaCl needed would approach 1180 nmol $m^{-2} s^{-1}$ (Yeo and Flowers, 1986). The same quantitative arguments apply as for loading the symplast initially and it does not appear credible that NaCl could be transported actively into the xylem.

6.2.4 *Regulation of uptake by the root*

Are there feedback effects from the shoot upon ion uptake in the roots? Does growth control ion uptake in the root or does ion transport from the root limit growth? These are questions of interest to physiologists as well as plant breeders. For a halophyte growing at high salinity it is arguable that NaCl is the principal 'nutrient': the plant cannot grow without osmotic adjustment and sodium and chloride are, by an order of magnitude, the most important constituents of the plant. As the external concentration of sodium chloride increases, so does ion content and growth (Figure 6.4). Ion content continues to increase up to 680 mol m^{-3} external NaCl but the rate of increase above about 100 mol m^{-3} is slow. Growth initially increases in line with ion content and then remains remarkably constant between about 100 and 350 mol m^{-3} external NaCl. Growth is inhibited below 100 mol m^{-3} and severely inhibited below about 20 mol m^{-3} (Figure 6.4). Growth also declines above about 350 mol m^{-3}. For the reasons discussed above it is legitimate to conclude that the capacity of the root to take up and transport ions is saturated at high concentrations of NaCl. This does not necessarily imply 'saturation' of carriers: the ATP resources of the cells may be used up in supporting the active transport of chloride and there is a *prima facie* case for regulation of net uptake so that the concentration in the root symplast remains sub-lethal. If ion transport across the root is proceeding at a maximal rate, then it is *not* being regulated by the shoot. There is over time, however, a close fit between net import of ions to the shoot and the

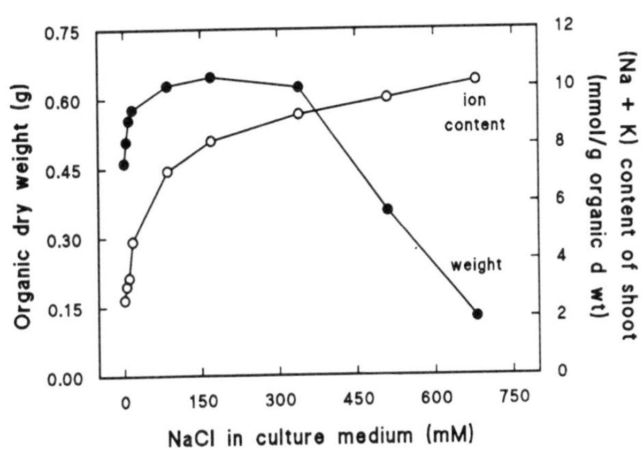

Figure 6.4 Growth and ion contents in the halophyte *Suaeda maritima* over a range of NaCl concentrations in the growth medium. Growth is measured as organic dry weight because inorganic ions represent a significant proportion of the dry weight. Drawn from data in Yeo and Flowers (1980; 1986).

relative growth rate of the shoot (Flowers and Yeo, 1988). While this is not evidence that feedback control does not occur, regulation of leaf growth by ion supply is a sufficient explanation. The relative constancy of leaf ion concentration over a range of external salinity suggests the plant operates at a much higher set-point than glycophytes and that the requirement of a high internal ion concentration is constitutive. This is equivalent to an average internal concentration of about 500 mol m^{-3} NaCl at the optimal growth range.

6.2.5 *Removal from the xylem*

All plants must balance the input of ions to their shoots and the capacity of the shoots to absorb or re-export those ions if toxic concentrations are not to accumulate in the leaf cells. Plants which are sensitive to salt in their leaves provide an example of selective removal of ions from the xylem. In a number of species, but particularly in maize (*Zea mays*) and beans (*Phaseolus vulgaris*), there is evidence for re-accumulation of sodium from the xylem. This is thought to be mediated by the xylem parenchyma cells which sometimes show anatomical changes consistent with enhanced rates of solute transport (cf. transfer cells, section 4.5.1). The capacity of such a system is, however, limited by the quantity of sodium that the plant can store in mature roots and petioles.

The balance between arrival of salt in the xylem and accumulation into the leaf cells is a fine one. The salt content in the leaf apoplast of *Suaeda maritima* is close to that in the vacuoles as deduced from the fact that turgor pressures in mature leaves are very low, despite the very low osmotic potentials found in this species (see Flowers *et al.*, 1986). The fluxes needed at the plasma membrane and tonoplast to explain the net ion accumulation in expanded cells in the leaf are unexceptional (Flowers and Yeo, 1986). How leaves accumulate ions does not pose comparable problems to those in the root, but the need for regulation of those fluxes is vital.

Salinity damage in salt-sensitive species provides an example of what happens if this fine balance between arrival in the xylem and accumulation by the cells of the leaf is not maintained. In rice (*Oryza sativa*), sodium chloride supplied to the roots arrives in the shoots via the xylem faster than the cells in the leaves can accumulate it. As a consequence the concentration of NaCl in the apoplast increases. This 'extracellular' salt accumulation was suggested by Oertli in 1968 as a cause of injury to plants, but direct evidence has only recently been available from X-ray microanalysis (Flowers *et al.*, 1991). The salt accumulation dehydrates the cells of the leaf leading to drought symptoms (the leaves roll) and appears to be the initialising event in salt damage in rice. The catastrophic consequences of this imbalance arise because the volume of the apoplast is small; about 1% of the solute available space in the leaf. Thus a quantity of solutes will change the concentration in the apoplast by about 100 times the concentration they would produce if accumulated into the vacuole.

6.2.6 *Compartmentation*

All cells compartmentalise ions, the vacuole is typically almost 2 pH units more acid than the cytoplasm (much more in CAM plants). However, few species appear able to compartmentalise NaCl. As well as compartmentation between protoplast and apoplast, salt tolerant plants compartmentalise salt between vacuole and cytoplasm. This is a necessary adaptation given the need for high salt concentrations in the plant to adjust the water potential to the very negative values of the environment and the toxicity of such high salt concentrations. In *S. maritima* the concentration of sodium chloride in the vacuole is typically 500 mol m^{-3} while in the cytoplasm it is about 150 mol m^{-3} and these conditions can be maintained for months (i.e. the life of the leaf). The requirements are for a transport system that accumulates NaCl into the vacuole, and for conditions that maintain this compartmentation at an affordable cost. If tonoplast membrane potentials are small then the free energy gradients for both Na^+ and Cl^- may be small, but the quantities are very large and hence the unit area flux needed to oppose leakage could potentially be large. In *S. maritima* several adaptations have been found in the lipid composition of the tonoplast that are consistent with minimising the passive leakage of ions across the lipid bilayer. This is not a pathway which would be expected to have a large conductance to NaCl so the existence of these adaptations emphasises the importance of maintaining compartmentation. The ion channels of higher plant tonoplasts appear to have large and poorly specific conductances at the voltages at which they are gated open. There is no evidence yet that the channels of halophytes differ from those of glycophytes in this respect. Since we can be fairly confident that these channels could almost never be open in a mature halophyte cell, because significant efflux of NaCl from the vacuole would be toxic, it is an intriguing question as to why plants appear to invest so heavily in tonoplast channels.

6.2.7 *Regulation of leaf concentration*

In *Suaeda maritima* it appears that new growth can utilise all the NaCl that the root system can supply but there has to be some provision for regulation in mature, expanded leaves. If salt arrives in the xylem it must be accumulated by leaf cells because the capacity for phloem export is strictly limited. Simple accumulation of the salt in the cells would increase the concentration of salt in the leaf and could exceed the concentration gradient that can be maintained across the tonoplast. The solution in *S. maritima* and some other halophytes is *succulence*. The volume of the mature cell increases with time, providing an increase in salt-storage capacity at constant concentration. Not all species have the ability to develop succulence.

In *S. maritima* any excess of xylem input of salt over that which can be absorbed by growth is small and succulence provides an adequate safety margin. However, this is not so for all halophytes. If salt cannot be accommodated by new growth and the limited capacities of retranslocation and succulence are exceeded, only one option can allow survival: excretion of excess salt. Excretion of salt via glands provides regulation of the leaf ion concentrations in almost complete independence from uptake at the root, since the glands have both large and variable capacity. There clearly need be no regulation of salt uptake by halophytes with salt glands, the excess is simply excreted.

6.2.8 *Salt glands*

Leaves are at the end of a delivery system of ions—the xylem containing the transpiration stream and its dissolved solutes. Where plants grow in high concentrations of dissolved salts some of these salts leak through into the xylem stream (the plant is imperfect in its ability to regulate this; see the discussion of bypass-flow in section 5.2). Consequently, if ions are continuously delivered in solution to the leaves as the water evaporates in transpiration, the ion concentration in the leaf will rise. It will rise to such an extent that the salts reach toxic concentrations, either through their osmotic effects or their effects on the metabolism. Ions might be removed from the leaves in the phloem (see chapter 7) but their fate then is to recirculate within the plant—this is not a solution to the problem of toxicity. Salt glands are able to secrete the excess salts arriving in the leaves through imperfect regulation of absorption. (see Flowers *et al.*, 1986).

Salt glands differ in structure between the families of flowering plants but all achieve a similar function, the secretion of salts from the leaves. Salt glands in the grasses are rather simple structures, ranging from the one-celled hairs of *Porteresia coarctata* (Figure 6.5) in the tribe Oryzeae, to the two-celled glands of the subfamily Chloridoideae (e.g. *Spartina* and *Cynodon*). In the dicotyledonous plants glands range in structure from the simple bladders present on the leaves of *Atriplex* species to much more complex structures (Figure 6.6) of between eight and forty cells; the latter occur on the leaves of *Tamarix* and of the mangrove species, *Aegilitis* (Fahn, 1988; Thomson *et al.*, 1988).

In spite of differences in complexity, salt glands from different species have many features in common. The cells that comprise the gland tend to be densely cytoplasmic with many mitochondria and often with proliferations of the plasma membrane (Thomson *et al.*, 1988). In some instances cell walls of the gland cells have the many protuberances characteristic of transfer cells. The outer walls of the gland cells subtending the space where salts accumulate (the *collecting compartment*) are lignified or impregnated with cuticle thus preventing an apoplastic continuity between these cells and the remainder of the leaf. There is, however, an apoplastic continuity between the basal cells of the gland and the remainder of the leaf (see below). Symplastic continuity between the gland cells themselves and the remainder of the leaf is maintained by the presence of numerous plasmodesmata.

The leaves of halophytes with salt glands can be readily observed to accumulate salt on their surfaces. The salts can be washed from the leaf and analysed. By floating leaves on solutions of variable composition or feeding

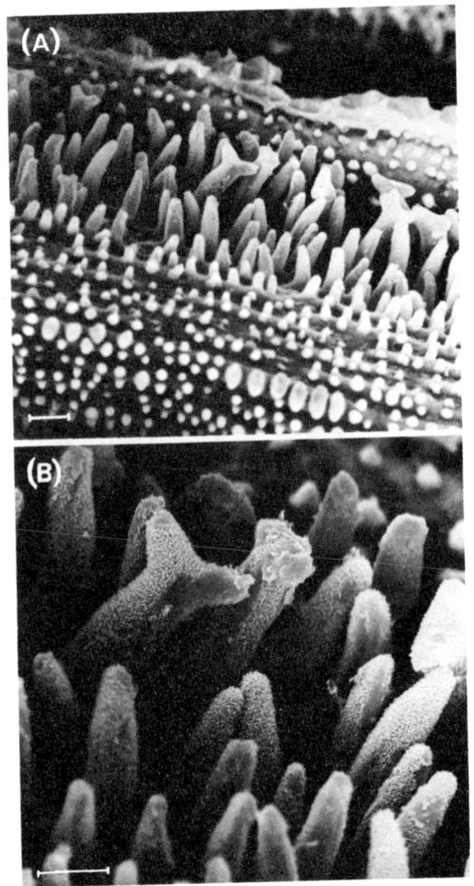

Figure 6.5 Scanning electron micrographs of part of a leaf from the tropical halophyte *Porteresia coarctata*. The micrographs show the many unicellular hairs or trichomes, which secrete salts and occur in furrows in the leaf surface. Magnification: top micrograph × 700; lower micrograph × 1900. Reproduced from Flowers *et al.* (1990).

salts through cut petioles, the capabilities of the glands for secretion can be evaluated. A great range of ions can be secreted, although those ions dominating the solution on which the leaf is floated tend to dominate the secretions.

Although there is evidence of selectivity in the secretion process, the selection may be conferred at various points on the pathway between root and shoot (for example, by withdrawal of potassium rather than sodium by

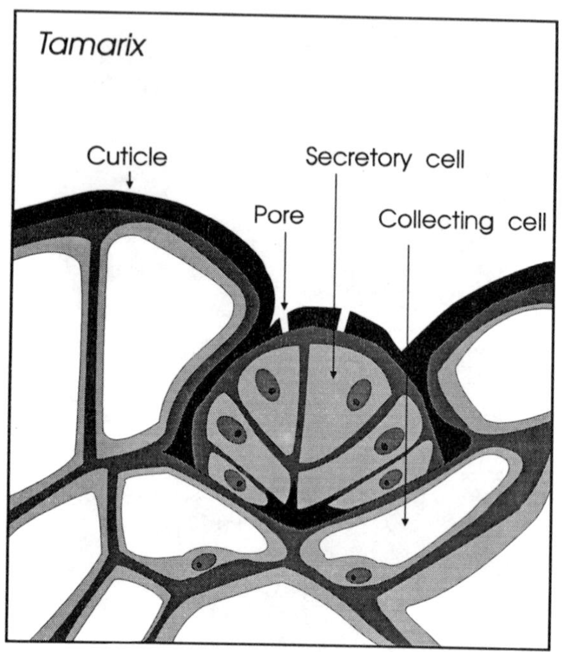

Figure 6.6 Diagrammatic representation of a longitudinal section through the multicellular salt gland of the leaf of *Tamarix*.

cells between root and shoot). The solution reaching the glands would be enriched in sodium relative to potassium, resulting in a secretion with a higher sodium to potassium ratio than the medium in which the plants were growing. Recent measurements on *Porteresia coarctata* show, however, that the glands do secrete more sodium than potassium in relation to the ion contents of the mesophyll cells (Table 6.2). Another important point is that a considerable proportion of the ions reaching the leaves are secreted (Table 6.3).

What then is the mechanism of this adjustment of the leaf's ion content? There appears to be an apoplastic continuum between the vascular tissues of leaves bearing salt glands and the glands themselves, and well developed symplastic connections between the gland cells and the underlying mesophyll (see Thomson *et al*., 1988). Experiments with detached leaves show clearly that the secretion process is temperature dependent and that anoxia and metabolic inhibitors reduce secretion: all these suggest an active process. Light also stimulates secretion. Secretion has been measured

Table 6.2 The ion concentrations in the vacuoles of mesophyll cells and of glands from the halophyte *Porteresia coarctata* determined by X-ray microanalysis. Note the much higher ratio of sodium to potassium in the gland cells, the contents of which are destined to be secreted. Data of Flowers *et al.* (1990)

NaCl in the culture solution ($mol\,m^{-3}$)	Element	Elemental concentration in the vacuoles of the cells ($mol\,m^{-3}$ analysed volume)		Na/K ratio	
		Mesophyll	Gland	Mesophyll	Gland
0	Na	5 ± 3	8 ± 4	0.08	1.3
	K	66 ± 16	6 ± 3		
	Cl	11 ± 5	$26 + 6$		
100	Na	79 ± 26	255 ± 39	0.9	7.3
	K	88 ± 18	35 ± 20		
	Cl	128 ± 37	203 ± 55		
200	Na	158 ± 39	323 ± 96	3.2	5.6
	K	50 ± 22	58 ± 19		
	Cl	172 ± 55	280 ± 78		

Table 6.3 The change in the ion contents of four separate leaves of the halophyte *Porteresia coarctata* expressed as μmol per leaf per day and the amount of salt washed from the leaves (i.e. the amount excreted per day). It is noticeable that the younger the leaf (leaf 1 is the oldest and 4 the youngest), the more salt it retains. The older leaves secrete more salt than they retain. Data of Flowers *et al.* (1990)

Leaf number	Change in content per day (μmol per leaf per day)		Gland activity as a percentage of a leaf accumulation
	Leaf	Washings	
1	39	80	205
2	167	141	84
3	280	183	65
4	244	104	43

against an electrochemical potential gradient in at least two species. Although the exact link between the light and secretion has not been established, the production of ATP during photosynthesis (see section 4.2) is an obvious candidate, given what is known of active transport in plants. It is not essential, however, that the ATP be used to drive transport in the gland cells themselves, since the glands (because they are part of the symplast) are coupled electrically to the cells of the mesophyll. The generation of a membrane potential in the mesophyll cells could be transmitted via the plasmodesmata to the gland cells, where, given the presence of the proper channels, ion efflux could be effected.

It is still not clear whether ion movement to the glands is symplastic apoplastic or a combination of the two. The latter seems the more likely of the possibilities as not only is there an apoplastic continuity between the xylem and the glands but there are also many plasmodesmatal connections between gland cells and the surrounding mesophyll. The fact that active transport occurs within the glands is supported by the dense cytoplasmic nature of some of the gland cells and the presence of numerous mitochondria. Ion fluxes have been calculated for various glands from measurements of the secretion rates, estimations of the number of glands and determinations of the amount of membrane material in a cell. The latter can be carried out by measurements made on electron micrographs (see Steer, 1981). Fluxes of Na^+ and Cl^- are very high (85 μmol $m^{-2} s^{-1}$) in *Limonium* in comparison to other measured fluxes across plant membranes (where high values are measured in terms of nmol $m^{-2} s^{-1}$; see section 6.2.2) if calculated for the surface area of the lateral walls of the gland (the transfusion zone: see Thomson *et al.*, 1988), but of average values if symplastic and expressed per plasmodesma (Na^+ 2.9×10^{-17} mol per plasmodesma per second; Cl^-, 1.4×10^{-18} mol per plasmodesma per second). A hypothesis that rationalises these measurements states that movement to the glands may occur through the apoplast or the symplast but entry into the gland is symplastic and it is in the glands themselves that the active transport concerned with secretion takes place.

In multicellular glands, the cell walls of the secretory cells are isolated from the remainder of the leaf apoplast by cuticularisation. This being so, secretion can take place into these cell walls without the possibility of the ions diffusing back into the remainder of the leaf apoplast. Water follows the salts osmotically and solution accumulates in the collecting compartment. Pressure builds in this compartment until the solution is discharged through surface pores. Once the solution has been discharged the process can begin again. Since large numbers of small vacuoles have been observed in the glands of some species (*Tamarix* and *Frankenia*) it is possible that secretion takes place, at least in these species, by a process of exocytosis (see section 4.3.1). Such a process would prevent high concentrations of ions, which can be toxic, from coming into contact with the cytoplasm.

CHAPTER SEVEN
TRANSPORT IN THE PHLOEM

The direction of the long distance transport of solutes in the xylem is governed by transpiration and so takes place from root to shoot. Transport in the reverse direction occurs in a specialised tissue known as *phloem* (from the Latin word for bark): together xylem and phloem make up the vascular tissue of the plant. The phloem is the means by which the products of photosynthesis are transported to non-photosynthetic parts of the plant, such as the roots, and to developing leaves, nectaries, fruits and seeds (Baker and Milburn, 1989). In contrast to the xylem, phloem is a tissue composed of *living* cells differentiated for the long distance transport of soluble organic compounds.

Experiments in the seventeenth and eighteenth centuries were the first to investigate what we now know as phloem transport. Such experiments, which can be readily repeated, involve the ringing of woody stems. A complete band of surface tissue is cut way or killed with steam around the stem or trunk. Tissue above the ring swells while that below eventually withers and dies. The inference to be drawn is that the transport of substances necessary for the sustenance of heterotrophic tissues is downward and takes place in the surface tissue of the stem. It was not, however, until the early years of the twentieth century that it was confirmed by chemical analysis that carbohydrates and nitrogen compounds move through layers of the bark containing the phloem (see Canny, 1973). Definitive evidence for transport in the cells of the phloem followed the availability of radionuclides in the 1940s and the use of microautoradiographic evidence from plants previously fed with $^{14}CO_2$. The isotope was shown to be in cells within the phloem. However, the question of what substances are naturally transported in the phloem has not been easy to answer.

7.1 Phloem sap

The main problem in determining what is transported in the phloem is how to obtain samples of sap for analysis. The sap from the phloem does not normally continue to flow if a cut is made into the tissue as most plants have a clotting mechanism that prevents continued bleeding from the phloem. Fortunately, however, there are ways in which sap can be obtained. Perhaps the simplest method, which is also the oldest, having been used by physiologists since the 1860s, is to make a cut into the bark (often of a tree; the sugar maple has been a popular subject). The cut must not be so deep that it penetrates the xylem or phloem exudate may be sucked into that tissue and lost. Exudation only takes place for a short time and the sap may be contaminated by the contents of any cell that is cut (see Canny, 1973; Ziegler, 1975). In a few species, notably *Ricinus communis* (the castor bean), *Lupinus albus* (the white lupin) and the flower stalk of species of *Yucca* (a genus of the Agavaceae), the flow continues over longer periods and substantial quantities of pure phloem sap can be obtained. Why in

Table 7.1 The composition of the phloem from a range of plant species. The absence of particular solutes is noted. Blanks in the table mean that information is not available. The data are from Hall and Baker (1972), Fukumorta and Chino (1982), Hayashi and Chino (1985) and Zimmerman and Milburn (1975).

Solute	Concentration (mol m^{-3})			
	Yucca flaccida	*Ricinus communis*	*Lupinus albus*	*Oryza sativa*
Sucrose	438–426	233–310	450	520–754
Amino acids	50–80[a]	37–4.2[a]	89[b]	270–600[a]
P_i	2.1	7.4–11.4		8.1
SO_4^{2-}	0.3	0.5–1.0		1.8
Cl^-	1	10–19		52
NO_3^-	absent	absent	absent	1.9
K^+	42	60–112	39	
Na^+	0.05	2.12	5	
Ca^{2+}	0.3	0.5–2.3	0.5	
Mg^{2+}	4.0	4.5–5		
NH_4^+		1.6		
ATP		0.4–0.6		
Auxin		0.60×10^{-4}		
GA		0.67×10^{-5}		
Cytokinin		0.52×10^{-4}		
ABA		0.40×10^{-3}		
pH	8.0–8.2		7.9	7.8–8.1
π^c (MPa)				−1.4
Viscosity (mPa s)		1.34		

[a]summation of individual concentrations.
[b]as glutamine.
[c]osmotic potential.

these few species, which have been important sources of information about phloem translocation, there is not a mechanism to prevent the leakage of phloem sap is unexplained. Another method for obtaining samples of pure phloem sap is to use insects such as aphids or planthoppers. These insects insert their stylets (mouth-parts) into the phloem and if the stylets are cut off, exudate continues to flow and may be collected for analysis (see for example, Fukumorita and Chino, 1982). Infestation by aphids and similar phloem-sucking insects is very common; presumably the small radius of the stylet allows sap to bleed without producing the pressure drop that would elicit the clotting response. Analysis of sap obtained by the variety of methods mentioned above has shown that sucrose is normally the dominant solute (Table 7.1). Its concentration in the phloem ranges from 200 to 900 $\mathrm{mol\,m^{-3}}$, with 400 to 500 $\mathrm{mol\,m^{-3}}$ as typical values. The concentration of sucrose taken from exudates from trees decreases with decreasing height by about 10 to 20 $\mathrm{mol\,m^{-3}\,m^{-1}}$ (see Canny, 1984). This is equivalent to a change in osmotic potential of 23 to 46 $\mathrm{kPa\,m^{-1}}$.

Table 7.2 Examples of common genera and the carbohydrates translocated in their phloem. Sucrose is present in the phloem sap of most plants but may be in combination with other sugars. Similarly raffinose, for example, may be present with stachyose or mannitol. For a comprehensive list see Appendix III in Zimmerman and Milburn (1975).

Sugar or sugar alcohol	Structure	Example of species translocating the sugar
Sucrose	Fru-Glu	Most plants
Raffinose	Fru-Glu-Gal	*Buddleia, Buxus, Fraxinus*
Stachyose	Fru-Glu-Gal-Gal	*Buddleia, Buxus, Fraxinus*
Verbascose	Fru-Glu-Gal-Gal-Gal	*Buddleia, Buxus, Fraxinus*
Mannitol	Hexitol	*Fraxinus*
Sorbitol	Hexitol	*Prunus, Pyrus, Sorbus*
Galacitol	Hexitol (Dulcitol)	*Euonymus*

glucose	fructose	galactose	mannitol	sorbitol	galacitol
CHO	CH_2OH	CHO	CH_2OH	CH_2OH	CH_2OH
HCOH	CO	HCOH	HOCH	HCOH	HCOH
HOCH	HOCH	HOCH	HOCH	HOCH	HOCH
HCOH	HCOH	HOCH	HCOH	HCOH	HOCH
HCOH	HCOH	HCOH	HCOH	HCOH	HCOH
CH_2OH	CH_2OH	CH_2OH	CH_2OH	CH_2OH	CH_2OH

Fru = fructose; Glu = glucose; Gal = galactose.

Although sucrose is the dominant sugar in many plants, it is not the only carbohydrate that is translocated. Oligosaccharides, particularly those in a family of compounds in which galactose units are added to sucrose (sugars in the raffinose series), and sugar alcohols are found in high concentrations in the phloem sap of some species (Table 7.2). The most common of these carbohydrates are raffinose itself, stachyose and the sugar alcohols, mannitol and sorbitol (see Ziegler, 1975). Reducing sugars are notably absent. Other important organic components in the phloem are aspartic and glutamic acids, asparagine and glutamine (nitrogenous compounds with high N:C ratios), ureides (allantoin and allantoic acid), as well as other amino acids and organic acids, hormones and ATP (Table 7.1). The dominant inorganic cation is potassium, present at concentrations of 40 to 100 mol m^{-3}, which presumably balances the charge on the dominant amino acids, aspartic and glutamic. The sap is a complex mixture of solutes with a pH of about 7.5 to 8 and an osmotic potential of between about −0.5 and −4.0 MPa; a very different solution from that flowing in the xylem (cf. Table 6.1).

7.2 Structure of the phloem

Phloem is often not easy to identify in transverse sections of plants but is generally spatially associated with the xylem; together they complete the vascular tissue of the plant. The phloem is composed of conducting elements, parenchyma cells, fibres and sclereids. Both the latter are supporting thickened cells (Behnke, 1989). Just as there are two sorts of conducting elements in the xylem, so the conducting elements of the phloem can be divided into two types. There are the relatively simple *sieve cells* (analagous to tracheids) found in the gymnosperms and *sieve tubes* (analagous to xylem vessels) in the angiosperms. The sieve tubes were discovered by Hartig in 1837; he also described exudation from the phloem in 1860. It is in the sieve cells and sieve tubes that solutes are translocated.

7.2.1 *Sieve tubes and companion cells*

Transport in the phloem in gymnosperms takes place through individual sieve cells. They are connected through sieve areas, which are areas of the cell wall with many cytoplasmic connections resembling plasmodesmata (see Fahn, 1990). In the angiosperms, however, the sieve tubes in which phloem transport occurs are formed of numerous cells. Just as the xylem vessels originate from a file of meristematic cells, so do the sieve tubes. They begin life as a chain of cells between which extensive intercellular connections develop during differentiation. The nucleus and cytoplasm become disorganised (see Cronshaw, 1981) and mature cells have neither nucleus, ribosomes nor tonoplast. In contrast to the xylem vessels, however, they retain a plasma membrane, an endoplasmic reticulum and mitochondria and in addition contain what is known as *p-protein* (phloem protein). The individual cells of the sieve tubes are known as *sieve tube elements* or *sieve elements* and their perforated end walls as *sieve plates* (Figure 7.1). A file of sieve elements make up a *sieve tube*. The other walls of the sieve elements are often thickened, having a shining appearance (termed nacreous) under the light microscope. About 20% of the area of the phloem is composed of sieve tubes which may be 100 to 500 μm in length and 20 to 30 μm in diameter (see Baker, 1978). The sieve plates are a particularly important component of the conduit as they are the connection between neighbouring elements.

The sieve plates connect sieve elements: a sieve plate is a common end-wall perforated by a large number of *sieve pores*. The pores are of greater diameter than plasmodesmata and without a desmotubule but lined with plasma membrane (Figure 7.1). They vary in diameter from less than 1 μm to about 14 μm (see Fahn, 1990); on average their diameter is 0.1 to 5 μm. The pores are about 1 μm long, the thickness of the sieve plate. They are surrounded by a polysaccharide, callose, similar to cellulose in being composed of glucose units, except that the units are linked by β-1,3 glycosidic bonds as opposed to the β-1,4 linkages in cellulose. Callose can develop over a matter of hours to occlude the pores completely.

Associated with the sieve tubes are cells that retain a dense cytoplasm and nucleus: these are the *companion cells* (see Figure 7.1). Companion cells arise from the same meristematic cell that gives rise to the sieve element, so the two cells are closely related ontogenetically. The meristematic cell divides longitudinally to form two cells. One becomes the sieve element and other the companion cell, or companion cells by transverse divisions. There is a variable number of companion cells associated with each sieve element, both between and within species. Companion cells contain dense cytoplasm, with many subcellular organelles such as mitochondria, and are interconnected by numerous plasmodesmata with the sieve elements. In the gymnosperms, the sieve cells are also associated with cells with dense cytoplasmic contents, known as *albuminous cells*. They presumably perform a similar function to the companion cells of the angiosperms. Albuminous cells and companion cells are presumed to control the metabolism of the sieve cells and sieve elements. In a few species (such as the broad bean, *Vicia*

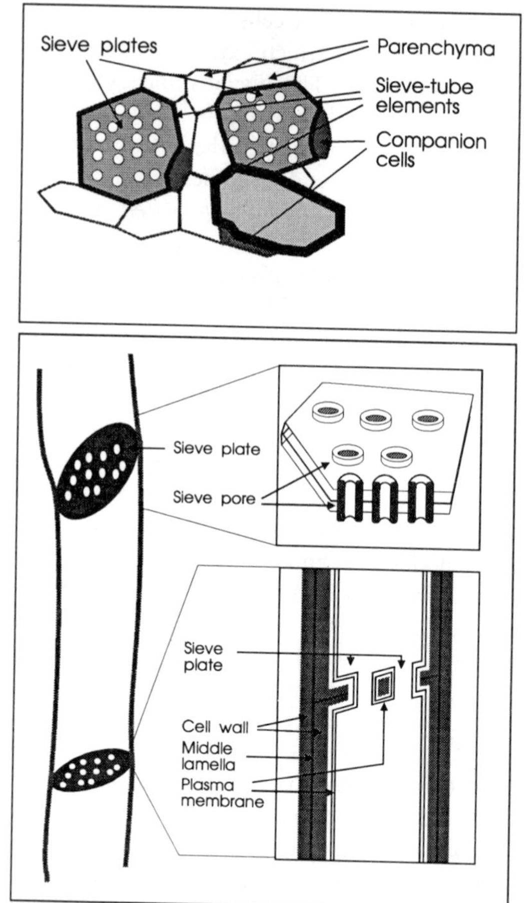

Figure 7.1 Structure of the phloem. The sieve tubes are composed of sieve elements separated by sieve plates. Sieve plates contain pores through which the phloem sap flows. The companion cells and phloem parenchyma make up the bulk of the remainder of the tissue.

faba), the companion cells have convoluted cells walls, such as are seen in some gland cells (see sections 4.5.1 and 6.2.8) and this is thought to be associated with increasing the surface area of the plasma membrane for transport.

If the sieve tubes are the conduits of the transport pathway, then it might, *a priori*, be expected that their lumina are sufficiently empty, except for the phloem sap that is being transported, to allow transport to occur. Similarly, the pores of the sieve plates might be expected to be free for the passage of solutes. Any impediment to flow would increase the driving force necessary to provide for translocation and so increase the resources that the plant has to commit. Surprisingly, many microscopists have reported blockages of various types within the sieve tubes and great controversy has surrounded the state of the sieve pores *in vivo* (Cronshaw, 1981). The explanation lies in the nature of the conducting elements. The sieve tubes, because they are living cells with a plasma membrane containing sugars in solution for transport, have a high turgor pressure (see above) so that once a sieve element is cut there is a release of pressure and the contents of the cell are forced towards the site of injury. A rapid response to such injury might be expected in a tissue that transports the valuable products of photosynthesis. This 'injury-induced response' disorganises the normal state of the cytoplasm within the sieve tube causing so-called 'slime plugs' containing p-protein to accumulate in the sieve plates blocking the pores. Consequently the structure of functional phloem cannot be inferred from micrographs of tissue obtained by simple cutting of samples for fixation. When particular care has been taken to prevent pressure release during the preparation of material for microscopy by, for example, prior freezing, pores do not appear blocked by p-protein (Cronshaw, 1981). The importance of open pores for transport can be affirmed by calculations of the resistance to flow through open and blocked pores. To make such calculations it is, however, necessary to have information on the rate of phloem transport.

7.2.2 *Rates of transport in the phloem*

It is only necessary to feed radiolabelled carbon (e.g. $^{14}CO_2$) to a single leaf and look for ^{14}C elsewhere in the plant to confirm that carbon is being translocated. ^{14}C is radioactive and decays with loss of a β-particle and so can be detected with a Geiger–Müller tube. By placing these counters at different distances from a leaf fed with $^{14}CO_2$, it is possible to time the arrival of counts at each detector and calculate the speed of translocation. The chief limitations to such a calculation are that any of the ^{14}C respired or moved laterally beyond the range of the counters will not be taken into account. With ^{14}C there is a particular problem of detection, as the β-particle emitted has a relatively low energy and is not easily measured *in vivo*. The technique only works at all with ^{14}C, because the phloem is a peripheral tissue. It is possible, however, to overcome the problem of detection by the use of another isotope, ^{11}C.

^{11}C emits positrons which are easy to count and it is therefore possible to detect much less carbon and that which lies deeper in the tissue than is possible with ^{14}C. However, ^{11}C has a very short half-life (20.3 minutes) for experimental purposes. This means that experiments have to be carried out near an accelerator suitable for producing the isotope. By recording the number of counts reaching a particular counter with time (Figure 7.2) the time taken for a fixed number of counts to reach each counter can be determined. By plotting this parameter against distance (the position of the counters) the velocity of transport can be calculated (Figure 7.2; see Moorby, 1981).

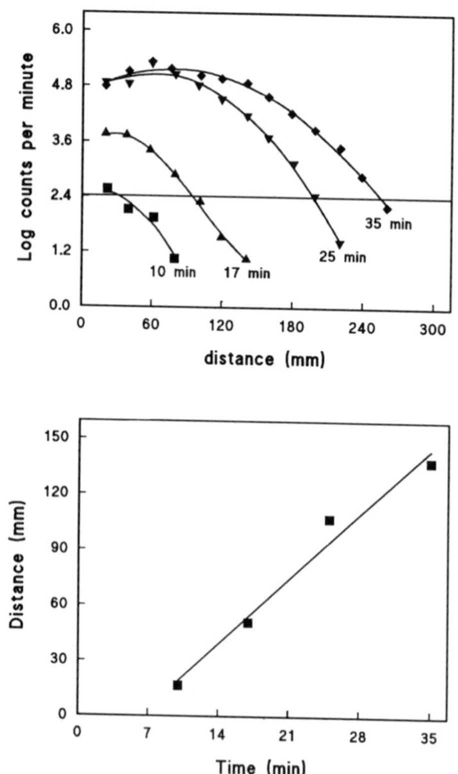

Figure 7.2 The movement of ^{14}C through the stem of a soybean (*Glycine max*). The top figure records the relationship between counts (on a logarithmic scale) and position on the stem at various times after feeding with $^{14}CO_2$. The lower graph shows the distance travelled by 250 counts with time. Redrawn from Moorby (1981).

Table 7.3 Estimates of the velocity of phloem transport. The data are from Canny (1973) and Moorby (1981).

Plant	Velocity ($mm\ min^{-1}$)
Willow (*Salix* sp.)	1–39
Ash (*Fraxinus americana*)	58–194
Wheat (*Triticum* sp.)	92–302
Sugar cane (*Saccharum officinarum*)	117–417
Maize (*Zea mays*)	430–580

Measurements of this type on a variety of plant species have determined the velocity of translocation generally to be between 133 and 267 $\mu m\ s^{-1}$ (8–19 $mm\ min^{-1}$ or 50–100 $cm\ h^{-1}$; Table 7.3). This is very much slower than transport in the xylem, but because the concentration of solutes in the phloem is much greater than in the xylem the quantities transported may be comparable. It appears that rates of phloem translocation in C_4 plants in general are faster than those in C_3 plants. The velocities can be converted to estimates of the amount of material being transported per unit time per unit area of phloem (specific mass transfer) from a knowledge of the contents of the phloem and the area of sieve tubes: specific mass transfer of a solution of 400 $mol\ m^{-3}$ moving at 170 $\mu m\ s^{-1}$ would be equivalent 68 mmol sucrose $s^{-1}\ m^{-2}$ phloem (245 $mol\ m^{-2}\ h^{-1}$ or 84 $kg\ m^{-2}\ h^{-1}$, both expressed per unit area of phloem). Specific mass transfer rates averaged over long times generally range from 10 to 50 $kg\ h^{-1}\ m^{-2}$ of phloem, but at peak rates may exceed 1000 $kg\ h^{-1}\ m^{-2}$ (Canny, 1984).

7.2.3 *Relationship between pore diameters and phloem transport*

The importance of pore diameters to flow through the sieve tubes and sieve plates can be evaluated with the Poiseuille equation (section 6.1.2):

$$J_v = -\frac{r^2 \Delta P}{8\eta l}$$

where J_v is the volume flowing per unit time per unit area in a cylinder of radius r and with a difference in pressure, ΔP, over a distance, l, driving the flow; η is the viscosity of the solution that is flowing. The negative sign is simply an indication that flow takes place in the direction of decreasing pressure. Since a volume flow with units of $m^3\ m^{-2}\ s^{-1}$ is formally equivalent to a velocity with units of $m\,s^{-1}$, it is possible to use the velocity measurements obtained from the feeding of radiolabelled carbon to calculate the pressure gradients required to drive flow through the cylinders that comprise the sieve elements. Similar calculations have been made to estimate the pressure required to drive phloem sap through the sieve plate pores and to evaluate the effects of occluding the pores with p-protein, when their diameter will be reduced.

The pressure gradient necessary to drive flow through a sieve element where the flow rate is $170\,\mu m\,s^{-1}$ of a solution with a viscocity (η) of 1.3 mPa s (1.34×10^{-3} for *Ricinus* phloem, see Table 7.1) and which has a radius (r) of $12\,\mu m$ and a length (l) of $500\,\mu m$ would be:

$$\Delta P \text{ per element} = \frac{-8\eta J_v}{r^2}\Delta l$$

$$= \frac{-8 \times 1.3 \times 10^{-3} \times 170 \times 10^{-6} \times 500 \times 10^{-6}}{(12 \times 10^{-6})^2}$$

$$= -6 \text{ Pa}$$

If the same flux were to cross a sieve plate, where the pores occupied half of the surface area, the flow rate through those pores would have to be double that through the lumen of the sieve element, viz. $340\,\mu m\,s^{-1}$. For pores with a radius (r') of $2\,\mu m$ and a length (l') of $1\,\mu m$, the pressure gradient would be:

$$\Delta P \text{ per sieve plate} = \frac{-8\eta J_v}{r'^2}\Delta l'$$

$$= \frac{-8 \times 1.3 \times 10^{-3} \times 340 \times 10^{-6} \times 1 \times 10^{-6}}{(2 \times 10^{-6})^2}$$

$$= -0.9 \text{ Pa.}$$

By having end walls set at an angle, the pore area presented per unit of area of the lumen is increased and therefore overall resistance to flow minimised. If the area available for flow was only 25% of that of the lumen, the pressure gradient required would be doubled.

To return to the original question of what happens if the pore is occluded with p-protein so that, say, only 10% of the area is available, then the pressure differences required would be:

$$\Delta P = \frac{-8 \times 1.3 \times 10^{-3} \times 340 \times 10^{-5} \times 1 \times 10^{-6}}{(0.2 \times 10^{-6})^2}$$

$$= -884 \text{ Pa.}$$

The pressure gradient across the pore is increased 1000-fold. It is clear that the ΔP necessary to drive flow very much depends on whether or not the pores are blocked. The geometry of the sieve pores does not, strictly, fulfil the conditions for which the Poiseuille equation is applicable but these calculations, which are commonly made, do indicate that a pressure-driven flow across a blocked pore would not be possible.

An explanation for the apparent contradiction between the open pores necessary for transport and the visible blockage of pores in many electron micrographs has already been made: it can be likened to bleeding in animals. The phloem, like the vascular system in animals, is under positive pressure (although the pressure may be 100 times higher in the phloem than in the arteries of mammals), so that damage to the sieve elements would lead to loss of the contents of the sieve cells or sieve tubes if there were not some mechanism to prevent this occurrence. Since plants have always been under threat of herbivory it is not surprising that a clotting mechanism evolved in the phloem, just as such a mechanism evolved to prevent loss of blood in animals.

The clotting mechanism is a two-stage process. Firstly there is the blocking of pores by the precipitation of p-protein, which occurs on the release of pressure. Secondly, there is a more permanent sealing of the pores by the synthesis of callose on the sieve plates. This is an enzymatic process and takes time, hence the need for the rapid response of precipitation of p-protein. This model of the behaviour of the phloem on injury provides an explanation for the structure of the phloem; it has a series of bulkheads (the sieve plates) which can be closed. The preparation of samples for electron microscopy is akin to herbivory and, unless specific precautions (such as prior rapid freezing) are taken, elicits the response to injury.

7.3 Mechanism of solute movement in the phloem

The rates of transport in the phloem (about 170 μm s^{-1}) are too fast, perhaps by some 10 000 times (see Canny, 1984), to occur by diffusion so some form of mass transport must, by definition, be involved. Mass flow of liquid through tubes is driven by gradients in pressure. For the phloem these gradients, calculated by application of the Poiseuille equation (see above), would be about 13 Pa per mm or 13 kPa m^{-1} if the pores in the sieve plates were open and the conditions otherwise as outlined above. The controversy over the state of the pores resulted, however, in a number of other hypotheses giving explanations of the translocation process (see chapters in Zimmermann and Milburn, 1975).

The discovery of p-protein in the sieve tube elements and sieve pores led to various suggestions that the protein itself was responsible for the motive force. It was assumed that the p-protein filaments had contractile properties similar to actin. However, although p-protein is present in most dicotyledonous plants that have been examined, it is absent from many palms, the duck weed *Lemna minor* and cereals such as maize, wheat and barley. Therefore, p-protein could not be part of any overall hypothesis to explain transport in the phloem. Furthermore, where p-protein is present it is not equivalent to actin or tubulin and, by 1981, Cronshaw concluded that this protein did not play a role in generating the force necessary for translocation. An alternative hypothesis suggested that water is induced to move through a charged pore in response to a gradient of electrical potential across the pore, the process being known as electro-osmosis. However, the fact that anions and cations are transported in the phloem was not explained by this electro-osmotic model. Furthermore, the calculated flux of potassium required to drive electro-osmosis is 10 000 times greater than that normally found for plant membrane systems and the energy requirements are inordinately large. This hypothesis is now redundant.

7.3.1 *Pressure flow*

As it is now accepted that the pores in the sieve plates are open in functional translocating phloem, a hypothesis developed before the controversy over p-protein regained general assent. The underlying principle of the hypothesis, published in the 1930s by Münch, is that the loading of solute into the phloem generates an osmotic gradient for water movement. Unloading of solute at a distant part of the pathway generates an osmotic gradient in the opposite direction. Water follows the solute into the sieve tubes where sucrose loading takes place, and out of the sieve tubes, where solute is unloaded. These two processes generate a pressure gradient, higher at the loading sites than at the sites where unloading occurs, which causes water and solutes to flow (Figure 7.3). The pressure gradients required ($13\,\mathrm{kPa\,m^{-1}}$, see above) to explain the measured velocities are quite within the bounds of possibility in plants. Estimates of the pressure within the sieve elements from the rates of flow from severed aphid stylets suggested values as high as 4 MPa in the sieve elements. Other estimates put values at between 0.7 and 1.2 MPa and the gradient at $20\,\mathrm{kPa\,m^{-1}}$ (see Canny, 1984). Significant amounts of energy are not required for the transport process *per se*, only for the loading and unloading processes. Furthermore, loading and unloading not only explain the mechanism of transport, but also the

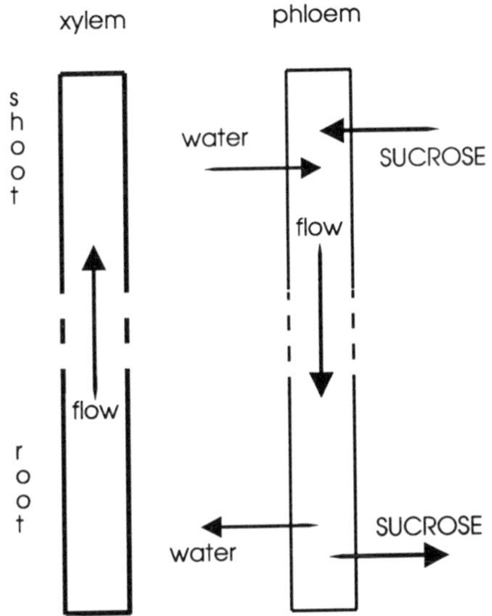

Figure 7.3 The pressure flow model (Münch hypothesis) of movement in the phloem and the relationship to transport in the xylem. Loading of sucrose into the phloem generates an osmotic gradient for the entry of water, which may be drawn from the xylem. Unloading reverses the gradient and leads to water transfer to the xylem.

direction of flow. Any part of the transport pathway from which solutes are unloaded will be a site to which transport takes place: it will be a *sink*. Similarly, any part of the phloem where solutes are being loaded will be a *source*. Transport can take place in opposite directions within the phloem of a given vascular bundle, but not within the same sieve tube. Young, expanding leaves, for example, requiring photosynthate for growth, will be a sink while expanded leaves producing sucrose are a source. Sinks, such as developing leaves, nectaries, roots and developing fruits, compete for the resources in the phloem.

7.4 Loading of solutes into the phloem

Any explanation of transport in the phloem must reconcile the loading of the major solutes (carbohydrates, commonly sucrose, amino acids and potassium) as well as the minor components (see Table 7.1) with the pH of the phloem sap (about 7.5 to 8.0) and the source of the materials being transported. These materials generally originate in chlorenchyma. Carbon is fixed during photosynthesis in the chloroplasts of the mesophyll cells; these chloroplasts are also the site of synthesis of amino acids. The chloroplasts are, therefore, the start of the transport pathway: the first barrier is the chloroplast envelope. Sucrose itself is synthesised by sucrose phosphate synthetase in the cytoplasm from the triose phosphate exported by the chloroplasts (see section 4.2). The next stage of the pathway is the movement from mesophyll cytoplasm to the phloem. Sucrose and amino acids must either pass through the symplast or diffuse by an apoplastic route to the sieve element/companion cell complex (or sieve cell/albuminous cell combination in gymnosperms), where loading into the transport pathway takes place. Evidence from a number of species suggests these solutes take a symplastic route through plasmodesmata to the sieve element/companion cell complex. For example, in wheat all longitudinal veins are surrounded by a structure analogous to the suberised layer in the endodermis known as the *mestome sheath*. It must presumably be crossed in the symplast. Calculations suggest the numbers of plasmodesmata are sufficient to transport sucrose at the required rates (see Giaguinta, 1983).

There is other evidence, however, that there is at least an apoplastic component to the path from mesophyll to phloem (Delrot, 1989). First, it can be demonstrated that radiolabelled sucrose can be loaded into the phloem if applied to a leaf surface and hence to the leaf apoplast. The label might be accumulated in the symplast, but at least there is the possibility of sucrose in the apoplast reaching the phloem. The route from chloroplast to sieve tube is not confined absolutely to the symplast. Second, the relatively membrane-impermeable compound *p*-chloromercuribenzene sulphonic acid (PCMBS) markedly and reversibly inhibits uptake and phloem loading of ^{14}C-sucrose in sugar beet leaves. In the short-term PCMBS does not affect photosynthesis or respiration. This is only explained simply by the PCMBS, which is assumed to remain in the apoplast, interfering with transfer of sucrose from apoplast to symplast at some stage of its journey. Third, sucrose is transported even in tissues where attempts have been made to disrupt plasmodesmata by prior plasmolysis. Fourth, it is not always clear that there are sufficient plasmodesmata to convey the necessary fluxes of solutes by the symplastic route alone although there clearly are in some cases (see Giaquinta, 1983). Finally, and perhaps most significantly, sucrose in the phloem may be more concentrated than in the mesophyll (see Giaquinta, 1983). This being so, symplastic transport alone is not a sufficient explanation since sucrose cannot diffuse against its concentration gradient, unless there is some as yet unknown mechanism to allow a change of concentration to be effected across plasmodesmata. The simplest hypothesis at present is that at some point, sugar is unloaded into the apoplast from where it is loaded back into the sieve elements by an active process. Even in species such as wheat, with its mestome sheath, sucrose presumably enters the vascular apoplast before being loaded into the sieve tubes.

The precise site at which this transference takes place remains equivocal. There is reason to suppose that unloading from the symplast occurs in the cells near the minor veins rather than throughout the mesophyll, since this will minimise both the length of diffusion pathways and interaction with water flow during transpiration. In some species, the presence of wall-bound invertases presents a problem to the apoplastic transport of sucrose: sucrose moving in the apoplast would presumably be hydrolysed to glucose and fructose.

The mechanism by which sucrose might enter the apoplast also remains obscure. Calculations of the transport of sucrose in broad bean (*Vicia faba*) suggest fluxes to be within the normal range for ion fluxes in plants (less than 100 nmol $m^{-2} s^{-1}$; Humphreys, 1988) and the presence of a channel through the plasma membrane that conducts sucrose from cytoplasm to cell wall has been proposed.

Over the past few years an increasing body of evidence has suggested that sucrose is loaded into the sieve elements by a proton symport from the neighbouring apoplast (see Giaquinta, 1983; Humphreys, 1988). Loading of sieve elements is most likely through the companion cells. Evidence for this is the similarity of the osmotic potential between companion cells and sieve elements, the fact that in some species the companion cells are modified as transfer cells and the presence of numerous and complex plasmodesmata between companion cells and sieve elements. The phloem membranes also show evidence of ATPase activity (see Cronshaw, 1981), which would be consistent with the presence of a proton pumping ATPase. It is, however, particularly difficult to evaluate the mechanism of solute loading by the sieve element/companion cell complex itself and so inference must be drawn from other systems in which sucrose is taken up by cells.

7.4.1 *Sucrose–proton cotransport*

Most information available for phloem loading relates to sucrose; presumably there are comparable specific mechanisms in those plants which translocate oligoasaccharides other than sucrose. There is now a substantial body of evidence to suggest that sucrose transport is coupled to the movement of a monovalent cation; to Na^+ in animals and to H^+ in plants and bacteria (see Reinhold and Kaplan, 1984). The carrier is a symporter and the gradient of the monovalent cation is generated in a primary active transport process. In plants the plasma membrane ATPase generates a gradient of protons (see section 3.2.2.1) and this in turn drives the uptake of sugar plus protons (Figure 7.4). The stoichiometry may be one proton per sugar transported (see Humphreys, 1988) while the driving force is the pmf, $\Delta\mu_{H^+}$ (see section 4.1) plus $+\Delta\mu_{sugar}$ (the sugar is accumulated into the phloem against its free energy gradient). The sugar molecule is, of course, neutral so its transport across the membrane in the presence of a proton must be electrogenic and will cause a transient depolarisation of the membrane potential. Depolarisation is transient because of homeostasis of the cytoplasmic proton concentration (pH), which means that protons are pumped out by the ATPase at an increased rate to compensate for the flux of protons through the symporter (Reinhold and Kaplan, 1984), There should also be a transient alkalinisation of the phase from which sucrose is

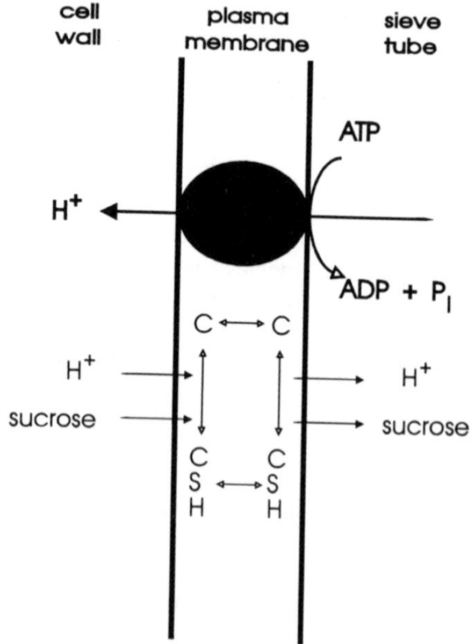

Figure 7.4 A model for the transport of sucrose into the sieve elements. ATPase activity at the plasma membrane generates a pmf and the return flow of protons is coupled to sucrose transport through a symporter. There are various suggestions about the detail of the reactions between the carrier and protons (see Humphreys, 1988) but at present they are a matter of speculation.

taken up, as protons and sugar cross the membrane into the cell. Interestingly, transport of sugars into the vacuole across the tonoplast cannot be a symport process, as protons are pumped into the vacuoles both by the tonoplast ATPase and the pyrophosphatase. In this case, if sucrose is to enter the vacuole by a mechanism coupled to the pmf, there must be a sucrose proton antiport.

It is difficult to obtain direct evidence of proton cotransport in the case of the phloem, as the phloem sieve elements and companion cells are such a small part of any tissue which is investigated. There is a very small ratio of phloem to non-phloem cells in most tissues. Using bark from willow in which a microelectrode was inserted into a severed aphid sylet, sucrose, but not mannitol, has been shown to induce a transient (20 min) depolarisation of the membrane potential (Wright and Fisher, 1981). Evidence of alkalinisation during sugar uptake has been obtained from experiments in which the vascular bundles of mature leaves of corn (*Zea mays*) are perfused with potassium chloride (5 mol m^{-3}, the sort of concentration that might be present in the xylem sap). The perfusate passes through the xylem and will also enter the apoplast of other cells in the vascular bundle, including the sieve element/companion cell complex, but not the mesophyll because the vascular tissue in this species is isolated from the mesophyll by suberised lamellae. As there are not plasmodesmatal connections between the sieve element/companion cell complex and the vascular parenchyma, any solute uptake must occur from the apoplast. Addition of sucrose (25 mol m^{-3}) to the perfusate causes a drop in the pH (by 0.6 to 0.9 pH units) of the solution emerging from the bundles, which lasts for about 30 to 60 minutes (Heyser, 1980).

The operation of proton cotransport is also consistent with the high pH of the phloem sap, the presence of ATP in the sap and the negative membrane potential normally maintained across the plasma membrane of the sieve elements (−150 mV; Wright and Fisher, 1981). The loading of sugar into the phloem is hypothesised to involve a carrier molecule that binds protons and sugar and it is the complex that crosses the membrane (Giaquinta, 1983; Humphreys, 1988; Figure 7.4). The order of binding of proton and sugar molecules and any charge carried by carrier or complex (which would alter their movement across the membrane because of the membrane potential) are all matters of speculation. The measured rates of phloem transport would require the apoplastic sucrose concentration to be about 20 mol m^{-3}.

Although carbohydrates such as sucrose are the main substances within the phloem of plants, potassium ions are also present at relatively high concentration (50–100 mol m^{-3}). How this potassium is involved in the transport process is not clear. It may be that transfer of potassium across the sieve element/companion cell complex simply balances the pmf against the available sucrose. In any event, potassium circulation in the phloem is substantial and important in the supply of this element to meristems (see below).

7.5 Solute unloading

Once in the sieve tube the sucrose generates an osmotic gradient for the entry of water. This, in turn, generates the pressure, which drives a mass flow of solution. Beyond the source, solutes must then be unloaded at any sink. Little is known of unloading. In some tissues, there are plasmodesmatal connections between phloem and the neighbouring cells of the sink that would allow symplastic transport. As long as sucrose concentrations are lower in the cells of the sink than those of the phloem and there are sufficient plasmodesmatal connections, sucrose movement can be by passive diffusion without the need to transfer the sucrose molecule across a membrane. If turgor were higher in the phloem than the sink, there might also be mass flow through the plasmodesmata and therefore turgor has been advocated a regulatory role in unloading (see Patrick, 1990). In other tissues, for example developing seeds, there must be apoplastic transport at some point in the pathway as there are no plasmodesmatal connections between maternal tissues and those of the developing seed. Unloading to the apoplast may occur by diffusion through the membrane of the companion cell or sieve element or through the putative sucrose/proton symporter responsible for loading the phloem (Patrick, 1990). There is evidence for a continuous leak and reloading during transport through the phloem. In areas where sucrose is required it may be hydrolysed by an acid invertase located in the apoplast, preventing the reloading. Overall, however, there is still only a very rudimentary understanding of the control of unloading.

7.6 Exchange of solutes between xylem and phloem

Although solutes such as sucrose are removed from the phloem and utilised in new growth, the requirement of growing cells for inorganic ions such as potassium and phosphorus appears to be less than the supply in the phloem. Consequently, there can be a significant recirculation of these ions within the xylem and the phloem (see Jeschke *et al.*, 1985). Potassium, for example, transferred from roots to shoots in the xylem, is loaded into the phloem and transferred back to the roots, where it leaks back into the xylem to circulate in the plant. Inorganic ions are not the only solutes to exchange between the phloem and the xylem; there can also be transfer of nitrogen compounds and of water itself (Pate, 1989). Although there can be downhill transport of sucrose from phloem to sink via the plasmodesmata, this is unlikely to be the case for any transfer of mineral elements from phloem to xylem. In this case, the ions, notably potassium, must move into the apoplast on their way to the xylem. The high solute concentration in the phloem, relative to the xylem, may lead to a lower water potential within the sieve tube elements than in the xylem elements in the leaves, in spite of the high turgor pressure in the phloem. There would be, in consequence, water flow from xylem to phloem. In the roots, the water potential gradient is likely to be in the opposite direction and water would flow from phloem to xylem (see Nobel, 1983).

There is also good evidence for the movement of substances from the xylem to the phloem. By comparing the ratios of a marker for xylem transport (inulin carboxylic acid) and an unnatural amino acid (α-aminoisobutyric acid) fed to the xylem of stems of tomato whose roots had been cut off, it has been shown that all the amino acid appearing in a mature leaf arrives via the xylem. At the apex, however, most of this amino acid arrives via the phloem (see Van Bel, 1990). There must be transfer from xylem to phloem. This is important in the overall physiology of the plant, as many amino acids are synthesised in the roots rather than the shoots and developing leaves in the apex requiring amino acids for growth are largely phloem fed. There is also autoradiographic evidence that assimilates can be transferred from phloem to xylem and that this can occur through parenchymal cells of the rays in woody plants (see Van Bel, 1990 for a detailed discussion of the pathway). Transport, which begins with diffusion from the xylem via a pit into the cell wall of a parenchymal cell may then be symplastic, as there are frequent plasmodesmata in the tangential walls of the ray cells, or apoplastic, as there are small radial canals formed in the intercellular spaces offering a potential route for solutes. It is likely that the symplastic route is more important. The rays cells have similar membrane potentials and as such constitute a symplast domain, i.e. a group of cells created by their isolation from other cells. Uptake of solutes by xylem parenchyma cells will occur by any of the various means already described that mediate such transport. Differences in water potential between xylem and phloem sap may also drive mass transfer from xylem to phloem.

REFERENCES

Chapter 1

Albert, R. and Kinzel, H. (1973) Unterscheidung von Physiotypen bei Halophyten des Neusiedlerseegebietes (Österreich). *Zeitschrift für Pflanzenphysiologie* **70**, 138–157.

Bental, M., Oren-Shamir, M., Avron, M. and Degani, H. (1988) ^{31}P and ^{13}C-NMR studies of the phosphorus and carbon metabolites in the halotolerant alga *Dunaliella salina*. *Plant Physiol.* **87**, 83–817.

Bowling, D.J.F. (1973) Measurement of a gradient of oxygen partial pressure across the intact root. *Planta* **111**, 323–328.

Bowling, D.J.F. (1987) Measurement of the apoplastic activity of K^+ and Cl^- in the leaf epidermis of *Commelina communis* in relation to stomatal activity. *J. Exp. Bot.* **38**, 1351–1355.

Bush, D.S. and Jones, R.L. (1990) Measuring intracellular Ca^{2+} levels in plant cells using the fluorescent probes, indo-1 and fura-2. *Plant Physiol.* **93**, 841–845.

Clarkson, D.T., Brownlee, C. and Ayling, S.M. (1988) Cytoplasmic calcium measurements in intact higher plant cells: results from fluorescence ratio imaging of fura-2. *J. Cell Sci.* **91**, 71–80.

Clipson, N.J.W., Hajibagheri, M.A. and Jennings, D.H. (1990) Ion compartmentation in the marine fungus *Dendryphiella salina* in response to salinity: X-ray microanalysis. *J. Exp. Bot.* **41**, 199–202.

Duncan, G. (1990) *Physics in the Life Sciences.* Blackwell, Oxford.

Felle, H. (1989) Ca^+-selective microelectrodes and their application to plant cells and tissues. *Plant Physiol.* **91**, 1239–1242.

Flowers, T.J. and Läuchli, A. (1983) Sodium versus potassium: substitution and compartmentation. In: *Inorganic Plant Nutrition* (Läuchli, A. and Pirson, A., eds). *Encyclopedia of Plant Physiology* **158**, Springer, Berlin, 651–681.

Flowers, T.J., Ward, M.E. and Hall, J.L. (1976) Salt tolerance in the halophyte *Suaeda maritima*: some properties of malate dehydrogenase. *Phil. Trans. Roy. Soc. London B* **273**, 523–540.

Hajibagheri, M.A. (1984) Physiological and ultrastructural studies of salt tolerance in the halophyte *Suaeda maritima* (L.) Dum. D.Phil. thesis, University of Sussex.

Hajibagheri, M.A., Yeo, A.R. and Flowers, T.J. (1985). Salt tolerance in the halophyte *Suaeda maritima* (L.) Dum. Fine structure and ion concentrations in the apical region of the roots. *New Phytol.* **99**, 331–343.

Hajibagheri, M.A., Flowers, T.J., Collins, J.C. and Yeo, A.R. (1988) A comparison of the methods of X-ray microanalysis, compartmental analysis and longitudinal ion profiles to estimate cytoplasmic ion concentrations in two maize varieties. *J. Exp. Bot.* **39**, 279–290.

Hall, T.A. and Gupta, B.L. (1983) The localization and assay of chemical elements by microprobe methods. *Quart. Rev. Biophys.* **16**, 279–339.

Hall, J.L., Harvey, D.M.R. and Flowers, T.J. (1978) Evidence for the cytoplasmic localization of betaine in leaf cells of *Suaeda maritima*. *Planta* **140**, 59–62.

Harvey, D.M.R., Flowers, T.J. and Hall, J.L. (1979) Precipitation procedures for sodium, potassium and chloride localisation in leaf cells of the halophyte *Suaeda maritima*. *J. Microscopy* **116**, 213–226.

Heinrich, G. (1989) Lamma ion spectra of the lattices of fungi. *J. Plant Physiol.* **133**, 770–772.

Huang, C.X. and van Steveninck, R.F.M. (1989) Longitudinal and transverse profiles of K^+ and Cl^- concentration in 'low'- and 'high-salt' barley roots. *New Phytol.* **112**, 475–480.

Jeschke, W.D. and Stelter, W. (1976) Measurement of longitudinal ion profiles in single roots of *Hordeum* and *Atriplex* by use of flameless atomic absorption spectroscopy. *Planta* **128**, 107–112.

Kinzel, H. (1982) *Pflanzenokologie und Mineralstoffwechsel*. Eugen Ulmer, Stuttgart.

Leigh, R.A. (1983) Methods, progress and potential for the use of isolated vacuoles in studies of solute transport in higher plant cells. *Physiol. Plant.* **57**, 390–396.

Leigh, R.A., Ahmad, N. and Wyn Jones, R.G. (1981) Assessment of glycinebetaine and proline compartmentation by analysis of isolated vacuoles. *Planta* **153**, 34–41.

Lüttge, U. and Weigl, J. (1965) Zur Mikroautoradiographie wasserlöslicher Substanzen. *Planta* **64**, 28–36.

Makjanic, J., Vis, R.D., Groneman, A.F., Gommers, F.J., and Henstra, S. (1988) Investigation of P and S distributions in the roots of *Tagetes patula* (L.) using micro-PIXIE. *J. Exp. Bot.* **39**, 1523–1528.

Malone, M., Leigh, R.A. and Tomos, A.D. (1989) Extraction and analysis of sap from individual wheat leaf cells: the effect of sampling speed on the osmotic pressure of extracted sap. *Plant, Cell and Environment* **12**, 919–926.

Mentré, P. and Escaig, F. (1988) Localization of cations by pyroantimonate. I. Influence of fixation on distribution of calcium and sodium: an approach by analytical ion microscopy. *J. Histochem. Cytochem.* **36**, 49–54.

Pahlich, E., Kerres, R. and Jäger, H.-J. (1983) Influence of water stress on the vacuole/extravacuole distribution of proline in protoplasts of *Nicotiana rustica*. *Plant Physiol.* **72**, 590–591.

Ratcliffe, R.G. (1987) Application of nuclear magnetic resonance methods to plant tissue. *Methods in Enzymology* **148**, 683–700.

Reid, R.J. and Smith, F.A. (1988) Measurement of the cytoplasmic pH of *Chara corallina* using double-barrelled pH micro-electrodes. *J. Exp. Bot.* **39**, 1421–1432.

Schaumann, L., Galle, P., Thelier, M. and Wissocq, J.C. (1988) Imaging the distribution of the stable isotopes of nitrogen ^{14}N and ^{15}N in biological samples by 'secondary ion emission microscopy'. *J. Histochem. Cytochem.* **36**, 37–39.

Stocking, C.R. (1956) Osmotic pressure or osmotic value. In: *Handbuch der Pflanzenphysiologie* (Ruhland, W., ed.). *III Allgemeine Physiologie der Pflanzenzelle*. Springer, Berlin, pp. 57–70.

Tsien, R.Y. and Poenie, M. (1986) Fluorescence ratio imaging: a new window into intracellular ionic signalling. *TIBS* **11**, 450–455.

Wyn Jones, R.G. and Pollard, A. (1983) Proteins, enzymes and inorganic ions. In: *Inorganic Plant Nutrition* (Läuchli, A. and Pirson, R.L., eds). Springer, Belin, pp. 528–562.

Zimmerman, U. and Steudle, E. (1980) Fundamental water relations parameters. In: *Plant Membrane Transport: Current Conceptual Issues* (Spanswick, R.M., Lucas, W.J. and Dainty, J., eds). Elsevier/North Holland Biomedical Press, pp. 113–127.

Chapter 2

Altmann, P.L. and Dittmer, D.S. (1966) *Environmental Biology Federation of American Societies for Experimental Biology*, Bethesda.

Asher, C.J. and Ozanne, P.G. (1967) Growth and potassium content of plants in solution cultures maintained at constant potassium concentrations. *Soil Science* **103**, 155–161.

Brown, D.J. and DuPort, F.M. (1989) Lipid composition of plasma membrane and endomembranes prepared from roots of barley (*Hordeum vulgare*, L.): effects of salt. *Plant Physiol.* **90**, 955–961.

Dainty, J. and Ferrier, J. (1989) Osmosis at the molecular level. *Studia Biophys.* **133**, 133–140.

Hall, J.L. (1983) Plasma membranes. In: *Isolation of Membranes and Organelles from Plant Cells* (Hall, J.L. and Moore, A.L., eds.). Academic Press, London, pp. 55–81.

Hall, J.L. and Baker, D.A. (1977) *Cell Membranes and Ion Transport.* Longman, London.

Hall, J.L. and Moore, A.L. (1983) *Isolation of Membranes and Organelles from Plant Cells.* Academic Press, London.

Hall, J.L., Flowers, T.J. and Roberts, R.M. (1982) *Plant Cell Structure and Metabolism.* Longman, London.

Haugland, R.P. (1989) Molecular probes. In: *Handbook of Fluorescence and Research Chemicals*, Molecular Probes Inc., Eugene, Oregon.

Hedrich, R. and Schroeder, J.I. (1989) The physiology of ion channels and electrogenic pumps in higher plants. *Ann. Rev. Plant Physiol.* **40**, 539–569.

Lannoye, R.J., Tarr, S.E. and Dainty, J. (1970) The effect of pH on the ionic and electrical properties of the internodal cells of *Chara australis. J. Exp. Bot.* **21**, 543–551.

Larsson, C. (1983) Partition in aqueous polymer two-phase systems: A rapid method for separation of membrane particles according to their surface properties. In: *Isolation of Membranes and Organelles from Plant Cells* (Hall, J.L. and Moore, A.L., eds). Academic Press, London, pp. 278–309.

Leigh, R.A. (1983) Methods, progress and potential for the use of isolated vacuoles in studies of solute transport in higher plant cells. *Physiol. Plant.* **57**, 390–396.

Lüttge, U. and Higinbotham, N. (1979) *Transport in Plants.* Springer-Verlag, New York.

Matthuius, F. (1991) Salt tolerance in *Plantago* and application of the patch-clamp technique in plant cell membranes, Rijks Universiteit, Groningen.

Nobel, P.S. (1974) *Biophysical Plant Physiology.* Freeman, San Fransisco.

Nobel, P.S. (1983) *Biophysical Plant Physiologyand Ecology.* Freeman, San Fransisco.

Robards, A.W. (1970) *Electron Microscopy and Plant Ultrastructure.* McGraw-Hill, London.

Robinson, R.A. and Stokes, R.H. (1959) *Electrolyte Solutions.* Butterworths, London.

Rochester, C.P., Kjellbom, P. and Larsson, C. (1987) Lipid composition of plasma membranes from barley leaves and roots, spinach leaves and cauliflower inflorescences. *Physiol. Plant.* **71**, 257–263.

Sandelius, A.S. and Morré, D.J. (1990) Plasma membrane isolation. In: *The Plant Plasma Membrane: Structure, Function and Molecular Biology* (Larsson, C. and Møller, I.M. eds). Springer-Verlag, Berlin, pp. 44–75.

Satter, R.L. and Moran, M. (1988) Ionic channels in plant cell membranes. *Physiol. Plant.* **72**, 816–820.

Slatyer, R.O. and Taylor, S.A. (1960) Terminology in plant and soil-water relations. *Nature* **187**, 922–924.

Starzak, M.E. (1984) *The Physical Chemistry of Membranes.* Academic Press, Orlando.

Takeda, K., Kurkdjian, A.C. and Kado, R.T. (1985) Ionic channels, ion transport and plant cell membranes: potential applications of the patch-clamp technique. *Protoplasma* **127**, 147–162.

Wagner, G. (1983) Higher plant vacuoles and tonoplasts. In: *Isolation of Membranes and Organelles from Plant Cells* (Hall, J.L. and Moore, A.L., eds). Academic Press, London, pp. 83–118.

Weast, R.C. (ed.) (1986) *CRC Handbook of Chemistry and Physics.* CRC Press, Boca Raton, Florida.

Yeagle, P. (1987) *The Membranes of Cells.* Academic Press, Orlando.

Yoshida, S. and Uemara, M. (1986) Lipid composition of plasma membrane and tonoplasts isolated from etiolated seedlings of mung bean (*Vigna radiate*, L.). *Plant Physiol.* **82**, 807–812.

Chapter 3

Balsamo, R.A. and Uribe, E.G. (1988) Plasmalemma-and tonoplast-ATPase activity in mesophyll protoplasts, vacuoles and microsomes of the Crassulacean-acid-metabolism plant *Kalanchoe daigremontiana*. *Planta* **173**, 190–196.

Bennett, A.B. and Spanswick, R.M. (1984) H^+-ATPase activity from storage tissue of *Beta vulgaris*: II H^+/ATP stoichiometry of an anion-sensitive H^+-ATPase. *Plant Physiol.* **74**, 545–548.

Bentrup, F.-W. (1990) Potassium ion channels in the plasmalemma *Physiol. Plant.* **79**, 705–711.

Benz, R. (1985) Porin from bacterial and mitochondrial outer membranes. *Crit. Rev. Biochem.* **19**, 145–190.

Briskin, D.P. (1990) The plasma membrane H^+-ATPase of higher plant cells: biochemistry and transport function. *Biochim. Biophys. Acta* **1019**, 95–109.

Crane, F.L. and Barr, R. (1989) Plasma membrane oxidoreductases. *Crit. Rev. Plant Sci.* **8**, 273–307.

Clarkson, D.T. and Hanson, J.B. (1980) The mineral nutrition of higher plants. *Ann. Rev. Plant Physiol.* **31**, 239–298.

Evans, D.E., Briars, S.A. and Williams, L.E. (1991) Active calcium transport by plant cell membranes. *J. Exp. Bot.* **42**, 285–303.

Ewing, N.N., Wimmers, L.E., Meyer, D.J., Chetelat, R.T. and Bennett, A.B. (1990) Molecular cloning of tomato plasma membrane H^+-ATPase. *Plant Physiol.* **94**, 1874–1881.

Giannini, J.L. and Briskin, D.P. (1987) Proton transport in plasma membrane and tonoplast vesicles from red beet (*Beta vulgaris*) storage tissue. *Plant Physiol.* **84**, 613–618.

Hanson, J.B. (1978) Application of the chemiosmotic hypothesis to ion transport across the root. *Plant Physiol.* **62**, 402–405.

Hedrich, R. and Schroeder, J.I. (1989) The physiology of ion channels and electrogenic pumps in higher plants. *Ann. Rev. Plant Physiol.* **40**, 539–569.

Henry, H. and Pilet, P.-M. (1988) Inhibition by vanadate of the tonoplast H^+ translocating ATPase of *Rubus* cells. *Plant Sci.* **56**, 149–154.

Hepler, P.K. and Wayne, R.O. (1985) Calcium and plant development. *Ann. Rev. Plant Physiol.* **36**, 397–439.

Johannes, E., Brosnan, J.M. and Sanders, D. (1991) Calcium channels and signal transduction in plant cells. *BioEssays* **13**, 331–336.

John, P. and Miller, A.J. (1986) Electrogenic proton translocation by the adenosine triphosphatase of intact vacuoles isolated from beet (*Beta vulgaris*). *J. Plant Physiol.* **122**, 1–16.

Lüttge, U. Higinbotham, N. (1979) *Transport in Plants*. Springer-Verlag, New York.

Lüttge, U. and Smith, J.A.C. (1988) CAM plants. In: *Solute Transport in Plant Cells and Tissues* (Baker, D.A. and Hall, J.L. eds). Longman Scientific and Technical, Harlow, pp. 417–452.

MacRobbie, E.A.C. (1988) Stomatal guard cells. In: *Solute Transport in Plant Cells and Tissues* (Baker, D.A. and Hall, J.L., eds). Longman Scientific and Technical, Harlow, pp. 453–497.

Mansfield, T.A., Hetherington, A.M. and Atkinson, C.J. (1990) Some current aspects of stomatal physiology. *Ann. Rev. Plant Physiol. Plant Mol. Biol.* **41**, 55–75.

Marschner, H. (1986) *Mineral Nutrition of Higher Plants*. Academic Press, London.

Mitchell, P. (1966) *Chemi-osmotic Coupling and Energy Transduction*. Glynn Research Ltd., Bodmin, Cornwall, England.

Møller, I.M. and Crane, F.L. (1990) Redox processes in the plasma membrane. In: *The Plant Plasma Membrane: Structure, Function and Molecular Biology* (Larsson, C. and Møller, I.M., eds). Springer-Verlag, Berlin, pp. 93–126.

Nakamoto, R.K. and Slayman, C.W. (1989) Molecular biology of the plasma-membrane [H^+]-ATPase of *Neurospora crassa*. In: *Plant Water Relations and Growth under Stress*

(Tazawa, M., Katsumi, M., Masuda, Y., and Okamoto, H., eds). Myu K.K., Tokyo, pp. 41–48.

Nelson, N. (1989) Structure, molecular genetics and evolution of vacuolar H^+-ATPases. *J. Bioenerg. Biomem.* **21**, 553–569.

Neilands, J.B. and Leong, S.A. (1986) Siderophores in relation to plant growth and disease. *Ann. Rev. Plant Physiol.* **37**, 187–208.

Poole, R.J. (1978) Energy coupling for membrane transport. *Ann. Rev. Plant Physiol.* **29**, 437–460.

Rayle, D.L. and Cleland, R.E. (1977) Control of plant cell enlargement by hydrogen ions. *Curr. Topics Develop. Biol.* **11**, 187–214.

Raven, J.A. (1988) Algae. In: *Solute Transport in Plant Cells and Tissues* (Baker, D.A. and Hall, J.L., eds). Longman Scientific and Technical, Harlow, pp. 166–219.

Rosen, B.P. (1986) Recent advances in bacterial ion transport. *Ann. Rev. Microbiol.* **40**, 263–286.

Sanders, D. (1988) Fungi. In: *Solute Transport in Plant Cells and Tissues* (Baker, D.A. and Hall, J.L., eds). Longman Scientific and Technical, Harlow, pp. 106–165.

Serrano, R. (1989) Structure and function of plasma membrane ATPase. *Ann. Rev. Plant Physiol. Plant Mol. Biol.* **40**, 61–94.

Serrano, R. (1990) Plasma Membrane ATPase. In: *The Plant Plasma Membrane: Structure, Function and Molecular Biology* (Larsson, C. and Møller, I.M., eds), Springer-Verlag, Berlin, pp. 127–153.

Spanswick, R.M. (1981) Electrogenic ion pumps. *Ann. Rev. Plant Physiol.* **32**, 267–289.

Stone, D.K. Crider, B.P., Südhof, T.C. and Xie, X.-S. (1989) Vacuolar proton pumps. *J. Bioenerg. Biomem.* **21**, 605–620.

Sussaman, M.R. and Harper, J.F. (1989) Molecular biology of the plasma membrane of higher plants. *The Plant Cell* **1**, 953–960.

Sze, H. (1985) H^+-translocating ATPase: advances using membrane vesicles. *Ann. Rev. Plant Physiol.* **36**, 175–208.

Tsien, R.W. and Tsien, R.Y. (1990) Calcium channels, stores and oscillations. *Ann. Rev. Cell Biol.* **6**, 715–760.

Taiz, L. (1986) Are biosynthetic reactions in plant cells thermodynamically coupled to glycolysis and the tonoplast proton motive force. *J. Theor. Biol.* **123**, 231–238.

Taiz, L., Gogarten, J.P. Kiback, H., Struve, I., Bernasconi, P., Rausch, T., Taix, S.L., Fichman, J. and Dittrich, P. (1989) Structure function and evolution of the vacuolar H^+-ATPase. In: *Plant Water Relations and Growth under Stress* (Tazawa, M., Katsumi, M., Masuda, Y. and Okamoto, H., eds). Myu K.K., Tokyo, pp. 57–64.

Tester, M. (1990) Plant ion channels: whole-cell and single-channel studies. *New Phytol.* **114**, 305–340.

Willmer, C.M. (1983) *Stomata*. Longman, London.

Chapter 4

Baskin, T.I. and Cande, W.Z. (1990) The structure and function of the mitotic spindle in flowering plants. *Ann. Rev. Plant Physiol. Mol. Biol.* **41**, 277–315.

Douce, R. and Neuburger, M. (1989) The uniqueness of plant mitochondria. *Ann. Rev. Plant Physiol. Plant Mol Biol.* **40**, 371–414.

Douce, R. and Day, D.A. (1985) *Higher Plant Cell Respiration*. Springer-Verlag, Berlin.

Dustin, P. (1984) *Microtubules*. Springer-Verlag, Berlin.

Fahn, A. (1979) *Secretory Tissues in Plants*. Academic Press, London.

Fahn, A. (1988) Secretory tissues in vascular plants. *New Phytol.* **108**, 229–257.

Findlay, N. (1988) Nectaries and other glands. In: *Solute Transport in Plant Cells and Tissues* (Baker, D.A. and Hall, J.L., eds). Longman Scientific and Technical Harlow, Essex, pp. 538–560.

Flowers, T.J. (1984) Chloride as nutrient and osmoticum. In: *Advances in Plant Nutrition, Vol 3* (Läuchli, A. and Tinker, B. eds). Praeger, New York, pp. 55–78.
Flügge, U.-I. and Heldt, H.W. (1991) Metabolite translocators of the chloroplast envelope. *Ann. Rev. Plant Physiol. Plant Mol. Biol.* **42**, 129–144.
Forte, M., Guy, H.R. and Manella, C.A. (1987) Molecular genetics of the VDAC ion channel: structural model and sequence analysis. *J. Bioenerg. Biomem.* **19**, 345–350.
Haliwell, B. (1984) *Chloroplast Metabolism: The Structure and Function of Chloroplasts in Green Leaf Cells.* Clarendon Press, Oxford.
Hall, J.L., Flowers, T.J. and Roberts, R.M. (1982) *Plant Cell Structure and Metabolism.* Longman, London.
Hanson, J.B. (1985) Membrane transport systems of plant mitochondria. In: *Higher Plant Cell Respiration* (Douce, R. and Day, D.A., eds). Springer-Verlag, Berlin, pp. 248–280.
Hedrich, R. and Schroeder, J.I. (1989) The physiology of ion channels and electrogenic pumps in higher plants. *Ann. Rev. Plant Physiol.* **40**, 539–569.
Heldt, H.W. and Flügge, U.-I. (1986) Transport of metabolites across the chloroplast envelope. *Methods in Enzymology* **125**, 705–716.
Kamiya, N. (1981) Physical and chemical basis of cytoplasmic streaming. *Ann. Rev. Plant Physiol.* **32**, 205–236.
Kuroda, K. (1990) Cytoplasmic streaming in plant cells. *Int. Rev. Cytology* **121**, 267–307.
Lloyd, C.W. (1987) The plant cytoskeleton: the impact of fluorescence microscopy. *Ann. Rev. Plant Physiol.* **38**, 119–39.
Manella, C.A. and Tedeschi, H. (1987) Importance of the mitochondrial outer membrane channel as a model biological channel. *J. Bioenerg. Biomem.* **19**, 305–308.
Margulis, L. (1981) *Symbiosis in Cell Evolution.* Freeman and Co., San Franscisco.
Meiners, S., Baron-Epel, O. and Schindler, M. (1988) Intercellular communication—filling in the gaps. *Plant Physiol.* **88**, 791–793.
Mitchell, P. (1966) *Chemiosmotic coupling in oxidative and photosynthetic phosphorylation.* Glenn Research Limited, Bodmin, Cornwall, England.
Nobel, R.S. (1984) *Biophysical Plant Physiology and Ecology.* W.H. Freeman, San Franscisco.
Prebble, J.N. (1988) Mitochondria and chloroplasts. In: *Solute Transport in Plant Cells and Tissues* (Baker, D.A. and Hall, J.L., eds). Longman Scientific and Technical, Harlow, Essex, pp. 28–82.
Robards, A.W. (1975) Plasmodesmata. *Ann. Rev. Plant Physiol.* **26**, 13–29.
Robards, A.W. and Lucas, W.J. (1990) Plasmodesmata. *Ann. Rev. Plant Physiol.* **41**, 369–419.
Robinson, D.G. and Depta, H. (1988) Coated vesicles. *Ann. Rev. Plant Physiol. Mol. Biol.* **39**, 53–99.
Robinson, D.G. and Hillmer, S. (1990) Endocytosis in plants. *Physiol. Plant.* **79**, 96–104.
Sivak, M.N., Leegood, R.C. and Walker, D.A. (1989) Transport of photoassimilates within photosynthetic cells. In: *Transport of Photoassimilates* (Baker, D.A. and Milburn, J.A., eds). Longman Scientific and Technical, Harlow, Essex, pp. 1–48.
Somerville, S.E. and Ogren, W.L. (1983) An *Arabidopsis thaliana* mutant defective in chloroplast dicarboxylate transport. *Proc. Nat. Acad. Sci. USA* **80**, 1290–1294.
Steer, M.W. (1988) Plasma membrane turnover in plant cells. *J. Exp. Bot.* **39**, 987–996.
Terry, B.R. and Robards, A.W. (1987) Hydrodynamic radius alone governs the mobility of molecules through plasmodesmata. *Planta* **171**, 145–157.

Chapter 5

Bowling, D.J.F. (1973) Measurement of a gradient of oxygen partial pressure across the intact root. *Planta* **111**, 323–328.
Bowling, D.J.F. (1981) Release of ions to the xylem in roots. *Physiol. Plant.* **53**, 392–397.
Boyer, J.S. (1985) Water transport. *Ann. Rev. Plant Physiol.* **36**, 473–516.
Briggs *et al.* (1961) In: *Electrolytes and Plant Cells.* Blackwell, Oxford.

Carpita, N., Sabularse, O., Montezinos, D. and Delmer, D.P. (1979) Determination of the pore size of cell walls of living plants. *Science* **105**, 1144–1147.

Clarkson, D.T. (1974) *Ion transport and cell structure in plants.* McGraw-Hill, London.

Clarkson, D.T. (1985) Factors affecting mineral nutrient acquisition by plants. *Ann. Rev. Plant Physiol.* **36**, 77–115.

Clarkson, D.T. (1988) Movements of ions across roots. In: *Solute Transport in Plant Cells and Tissues.* (Baker, D.A. and Hall, J.L., eds), Longman Scientific and Technical, Harlow

Crafts, A.S. and Broyer, T.C. (1938) Migration of salts and water into xylem of roots of higher plants. *Am. J. Bot.* **25**, 529–535.

Drew, M.C. and Saker, L.R. (1986) Transport to the xylem in aerenchymatous roots of *Zea mays J. Exp. Bot.* **37**, 22–33.

Epstein, E. (1972) *Mineral Nutrition of Plants: Principles and Perspectives.* John Wiley and Sons, Inc., New York.

Glass, A.D.M. and Perley, J.E. (1979) Cytoplasmic streaming in the root cortex and its role in the delivery of potassium to the shoot. *Planta* **145**, 399–401.

Leigh, R.A. and Wyn Jones, R.G. (1984) A hypothesis relating critical potassium concentration for growth to the distribution and functions of this ion in the plant cell. *New Phytol.* **97**, 1–13.

Lüttge, U. and Higinbotham, N. (1979) *Transport in Plants.* Springer-Verlag, New York.

Nissen, P. (1991) Multiphasic uptake mechanisms in plants. *Internat. Rev. Cytology* **126**, 89–134.

Nye, P.H. and Tinker, P.B. (1977) *Solute Movement in the Soil-Root System.* Blackwell, Oxford.

Peterson, C.A. (1988) Exodermal Casparian bands: their significance for ion uptake roots. *Physiol. Plant.* **72**, 204.

Preston, R.D. (1974) *The Physical Biology of Plant Cell Walls.* Chapman and Hall, London.

Sanders, F.E.T. and Tinker, R.B. (1971) Mechanism of absorption of phosphate from the soil by *Endogone mycorrhzas. Nature* **223**, 278–279.

Steudle, E. and Jeschke, W.D. (1983) Water transport in barley roots: measurements of root pressure and hydraulic conductivity of roots in parallel with turgor and hydraulic conductivity of root cells. *Planta* **158**, 237–248.

Taiz, L. and Zeiger, E. (1991) *Plant Physiology.* Benjamin/Cummings Publishing Company, Inc. Redwood City, Ca, USA.

Tyree, M.Y. (1970) The symplast concept; a general theory of symplastic transport according to the thermodynamics of irreversible processes. J. *Theoret. Biol.* **26**, 181–214.

Wyn Jones, R.G. (1975) Excised roots. In: *Ion Transport in Plant Cells and Tissues.* (Baker, D.A. and Hall, J.L., eds). North-Holland, Amsterdam, pp. 193–229.

Chapter 6

Boyer, J.S. (1985) Water transport. *Ann. Rev. Plant Physiol.* **36**, 473–516.

Canny, M.J. (1990) What becomes of the transpiration stream? *New Phytol.* **114**, 341–368.

Fahn, A. (1988) Secretory tissues in vascular plants. *New Phytol.* **108**, 229–257.

Fahn, A. (1990) *Plant Anatomy.* Pergammon Press, Oxford.

Farquhar, G.D. and Raschke, K. (1978) On the resistance to transpiration of the sites of evaporation within the leaf *Plant Physiol.* **61**, 1000–1005.

Flowers, T.J. and Yeo, A.R. (1988) Ion relations of salt tolerance. In: *Solute Transport in Plant Cells and Tissues* (Baker, D.A. and Hall, J.L. eds). Longman Scientific and Technical, Harlow, pp. 392–416.

Flowers, T.J., Hagibagheri, M.A. and Clipson, N.J.W. (1986) Halophytes. *Quart. Rev. Biol.* **61**, 313–337.

Flowers, T.J. Hajibagheri, M.A. and Yeo, A.R. (1991) Ion accumulation in the cell walls of rice plants growing under saline conditions: evidence for the Oertli hypothesis. *Plant Cell and Environment* **14**, 319–325.

Flowers, T.J., Flowers, S.A., Hajibagheri, M.A. and Yeo, A.R. (1990) Salt tolerance in the halophytic wild rice *Porteresia coarctata* Tateoka. *New Phytol.* **114**, 675–684.
Nobel, P.S. (1983) *Biophysical Plant Physiology and Ecology*. Freeman, San Fransisco.
Oertli, J.J. (1968) Extracellular salt accumulation, a possible mechanism of salt injury in plants. *Agrochimica* **12**, 461–469.
Passioura, J.B. (1988) Movement of water in and to roots *Ann. Rev. Plant Physiol.* **39**, 245–265.
Schurr, U. and Gollan, T. (1990) Composition of xylen sap of plants experiencing root water stress—a descriptive study. In: *Importance of root to shoot communication in the responses to environmental stress* (Davies, W.J. and Jeffcoat, E., eds). British Society for Plant Growth Regulation, Bristol, pp. 201–214.
Steer, M.W. (1981) Understanding Cell Structure. Cambridge University Press, Cambridge.
Taiz, L. and Zeiger, E. (1991) *Plant Physiology*. The Benjamin/Cummings Publishing Co. Inc., Redwood City.
Thomson, W.W., Faraday, C.D., and Oross, J.W. (1988) Salt glands. In: *Solute Transport in Plant Cells and Tissues* (Baker, D.A. and Hall, J.L. eds). Longman Scientific and Technical, pp. 498–537.
Yeo, A.R. and Flowers, T.J. (1980) Salt tolerance in the halophyte *Suaeda maritima*: evaluation of the effect of salinity upon growth. *J. Exp. Bot.* **31**, 1171–1183.
Yeo, A.R. and Flowers, T.J. (1986) Ion transport in *Suaeda maritima*: its relation to growth and implications for the pathway of radial transport of ions across the root. *J. Exp. Bot.* **37**, 143–159.

Chapter 7

Baker, D.A. (1978) *Transport Phenomena in Plants*. Chapman and Hall, London.
Baker, D.A. and Milburn, J.A (1989) *Transport of Photoassimilates*. Longman Scientific and Technical, Harlow, Essex.
Behnke, H-D. (1989) Structure of the phloem. In: *Transport of Photoassimilates* (Baker, D.A. and Milburn, J.A. eds). Longman Scientific and Technical, Harlow, Essex, pp. 79–137.
Canny, M.J. (1973) *Phloem Translocation*, Cambridge University Press, Cambridge.
Canny, M.J. (1984) Translocation of nutrients and hormones. In: *Advanced Plant Physiology* (Wilkins, M.J. ed.) Pitman, London, pp. 277–296.
Cronshaw, J. (1981) Phloem structure and function. *Ann. Rev. Plant Physiol.* **32**, 465–484.
Delrot, S. (1989) Loading of photoassimilates. In: *Transport of Photoassimilates* (Baker, D.A. and Milburn, J.A. eds). Longman Scientific and Technical, Harlow, Essex, pp. 167–206.
Fahn, A. (1990) *Plant Anatomy*. Pergamon Press, Oxford.
Fukumorita, T. and Chino, M. (1982) Sugar, amino acid and inorganic contents in rich phloem sap. *Plant Cell Physiol.* **23**, 273–283.
Giaquinta, R.T. (1983) Phloem loading of sucrose. *Ann. Rev. Plant Physiol.* **34**, 347–387.
Hall, S.M. and Baker, D.A. (1972) The chemical composition of *Ricinus* phloem exudate. *Planta* **106**, 131–140.
Hayashi, H. and Chino, M. (1986) Collection of pure phloem sap from wheat and its chemical composition. *Plant Cell Physiol.* **26**, 325–330.
Heyser, W. (1980) Phloem loading in the maize leaf. *Ber. deutsch. bot. Gesell.* **93**, 221–228.
Humphreys, T.E. (1988) Phloem transport-with emphasis on loading and unloading. In: *Solute Transport in Plant Cells and Tissues* (Baker, D.A. and Hall, J.L. eds). Longman Scientific and Technical, Harlow, Essex, pp. 305–345.
Jeschke, W.D., Atkins, C.A. and Pate, J.S. (1985) Ion circulation via phloem and xylem between root and shoot of nodulated white lupin. *J. Plant Physiol.* **117**, 319–330.
Moorby, J. (1981) *Transport Systems in Plants*. Longman. London.
Münch, E. (1930) Die Stoffbewegungenin der Pflanze. Fischer, Jena.
Nobel, P.S. (1983) *Biophysical Plant Physiology and Ecology*. Freeman, San Fransisco.

Pate, J.S. (1989) Origin, destination and fate of phloem solutes in relation to organ and whole plant functioning. In: *Transport of Photoassimilates* (Baker, D.A. and Milburn, J.A. eds). Longman Scientific and Technical, Harlow, Essex, pp. 138–166.

Patrick, J.W. (1990) Sieve element unloading: cellular pathway, mechanism and control. *Physiol. Plant.* **78**, 298–308.

Reinhold, L. and Kaplan, A. (1984) Membrane transport of sugars and amino acids. *Ann. Rev. Plant Physiol.* **35**, 45–83.

Van Bel, A.J.E. (1990) Xylem-phloem exchange via the rays: the undervalued route of transport. *J. Exp. Bot.* **41**, 631–644.

Wright, J.P. and Fisher, D.B. (1981) Measurement of the sieve tube membrane potential. *Plant Physiol.* **67**, 845–848.

Ziegler, H. (1975) Nature of transported substances, In: *Transport in Plants: I Phloem Transport* (Zimmerman, M.H. and Milburn, J.A., eds). *Encyclopedia of Plant Physiology* (New series) Volume 1. Springer-Verlag, Berlin, pp. 59–100.

Zimmerman, M.H. and Milburn, J.A. (1975) *Encyclopedia of Plant Physiology* (New series) Volume 1. Springer-Verlag, Berlin.

Index